# Sustainable Development Goals Series

The **Sustainable Development Goals Series** is Springer Nature's inaugural cross-imprint book series that addresses and supports the United Nations' seventeen Sustainable Development Goals. The series fosters comprehensive research focused on these global targets and endeavours to address some of society's greatest grand challenges. The SDGs are inherently multidisciplinary, and they bring people working across different fields together and working towards a common goal. In this spirit, the Sustainable Development Goals series is the first at Springer Nature to publish books under both the Springer and Palgrave Macmillan imprints, bringing the strengths of our imprints together.

The Sustainable Development Goals Series is organized into eighteen subseries: one subseries based around each of the seventeen respective Sustainable Development Goals, and an eighteenth subseries, "Connecting the Goals," which serves as a home for volumes addressing multiple goals or studying the SDGs as a whole. Each subseries is guided by an expert Subseries Advisor with years or decades of experience studying and addressing core components of their respective Goal.

The SDG Series has a remit as broad as the SDGs themselves, and contributions are welcome from scientists, academics, policymakers, and researchers working in fields related to any of the seventeen goals. If you are interested in contributing a monograph or curated volume to the series, please contact the Publishers: Zachary Romano [Springer; zachary.romano@springer.com] and Rachael Ballard [Palgrave Macmillan; rachael.ballard@palgrave.com].

Thomas Walker · Carmela Cucuzzella ·
Sherif Goubran · Rana Geith
Editors

# The Role of Design, Construction, and Real Estate in Advancing the Sustainable Development Goals

palgrave
macmillan

*Editors*
Thomas Walker
Department of Finance
Concordia University
Montreal, QC, Canada

Carmela Cucuzzella
Faculty of Environmental Design
Université de Montréal
Montreal, QC, Canada

Sherif Goubran
Department of Architecture
The American University in Cairo
New Cairo, Egypt

Rana Geith
Department of Architecture
The American University in Cairo
New Cairo, Egypt

ISSN 2523-3084 ISSN 2523-3092 (electronic)
Sustainable Development Goals Series
ISBN 978-3-031-28738-1 ISBN 978-3-031-28739-8 (eBook)
https://doi.org/10.1007/978-3-031-28739-8



This Palgrave Macmillan imprint is published by the registered company Springer Nature Switzerland AG
The registered company address is: Gewerbestrasse 11, 6330 Cham, Switzerland

# Preface

The publication of the 2030 Agenda signaled an urgent call for transformative change toward a more sustainable world: one that is resource-efficient, economically conscious, and socially driven. Scholars have been studying how the building and construction industries can help advance the Sustainable Development Goals (SDGs) to move beyond the traditional environmental risk management approaches, revealing complex challenges and opportunities across the lifecycle of building projects. Prior publications have frequently followed thematic or disciplinary approaches focusing on specific building project phases. However, this resulted in a lack of integrative research and application to study the impacts of buildings on the environment, societies, and economies within the built environment more extensively.

This book expands the reader's understanding of sustainability in the built environment beyond the design and construction phases by collecting insights from scholars and practitioners to uncover the role the design, construction, and real estate sectors play in advancing the SDGs across the complete lifecycle of building projects, namely the built environment's design, construction, investment, management, and regulatory dimensions. The book prioritizes schemes that follow a lifecycle-based approach to the topic, addressing design, construction, management, investment, and regulatory dimensions of projects in the area. The collection explores interconnections between different project phases to provide insights into the multi dimensional and transformative approaches that encompass the interconnected nature of the SDGs and emphasizes the importance of multi-stakeholder synergies in attaining them.

The editors invited contributions from the international community of scholars and practitioners at the interface of architecture, construction, real estate investment and portfolio management, policymaking, and sustainable development.

The editors accepted contributions that are transdisciplinary in their approach, and strive to incorporate new concepts or tools that move beyond the boundaries of academics. The editors also encouraged contributions incorporating concepts from fields traditionally not at the core of the academic discourse on the built environment, such as the natural and social sciences and the arts. Most importantly, the editors prioritized contributions that move beyond encouraging eco-efficiencies and incremental improvements in current practices, presenting experimental approaches or concepts yet to be applied or scaled.

The resulting collection provides a snapshot of current and potential interlinkages between the building sector and the 2030 Agenda, highlighting the key challenges that require the urgent attention of researchers, businesses, and policymakers. These include, but are not limited to, the gaps in current industry practices, the limits of current assessment schemes in advancing the 2030 Agenda, inconsistencies between policies and regulations and the SDGs, and the potential of new methodologies, approaches, and techniques in addressing the urgent development challenges of the built environment. Thus, the book serves as a critical reference for built environment scholars, practitioners, and policymakers aiming to embed the SDGs in their respective work.

Montreal, Canada Thomas Walker
Montreal, Canada Carmela Cucuzzella
Cairo, Egypt Sherif Goubran
Cairo, Egypt Rana Geith

# Acknowledgments

We acknowledge the financial support provided through the Jacques Ménard—BMO Centre for Capital Markets at Concordia University in Montreal, the Office of the Associate Provost for Research Innovation and Creativity at the American University in Cairo, and the Social Sciences and Humanities Research Council (SSHRC). We would also like to thank Concordia University and the American University in Cairo for the academic support they provided. In addition, we appreciate the excellent copy-editing and editorial assistance we received from Aya Elshabshiri, Eimear Rosato, Laila El-Refai, Miles Murphy, Virginia Bassily, and Victoria Kelly. Finally, we would like to thank all the authors of the chapters for their diverse and insightful contributions.

# Contents

# Notes on Contributors

**Alpay Akgüç** received his B.Sc. in Mechanical Engineering from Bülent Ecevit University in 2007. Following his graduation from the Heat-Fluids Postgraduate Program at the Faculty of Mechanical Engineering at Istanbul Technical University (ITU) in 2010, he worked as a building performance modeling specialist at the International Center for Applied Thermodynamics (ICAT) of Yeditepe University. In 2011, Dr. Akgüç was admitted to the Building Sciences Ph.D. Program at the Faculty of Architecture, ITU, and assumed the position of project manager at Ecological Architectural Consultancy (EKOMIM), an R&D firm at ITU Ari Teknokent. Thereafter, he focused on the assessment of the energy performance of buildings in the scope of green building certification systems, accommodating building simulation tools that used detailed-dynamic calculation methods. He also worked as a Green Building Specialist, during which time he oversaw green building certification systems. Following his successful completion of a Ph.D. degree in 2019, Dr. Akgüç became a lecturer at the Department of Architecture of Girne American University, where he focused on physical environmental control and building energy efficiency. He is currently an Assistant Professor in the Faculty of Architecture in Istanbul Aydın University and publishes scientific studies on his ongoing research in the fields of building energy efficiency.

**Dheeraj Arrabothu** works for the Global Green Growth Institute on promoting green building and construction practices in Rwanda. He has worked on developing the Rwanda Green Building Minimum Compliance System. Furthermore, he has conducted Rapid building performance audits of existing public buildings and has conducted research on circular built environment practices in Rwanda, while supporting the functioning of the Rwanda Green Building Organization. He has also worked for the Indian Green Building Council, where he developed green cities and green mass transit rating systems. He has published works with a focus on green materials and green buildings in developing countries.

**Claire Brown** is a Ph.D. researcher at Tyndall Centre for Climate Change at the University of Manchester. Claire's research focuses on addressing issues around heating and cooling demand in social housing in the UK. She examines how new-build social housing could be a potential way to reduce demand on the power grid through better design, as well as further opportunities for self-generation. She also explores how commercially viable solutions might exist to allow social housing to be built with high-energy performance built-in. She has published in The Conversation and a blog for the Social Policy Association, as well as the University of Manchester's significant publication, "Building Utopia," which focusing on what is needed for significant policy change. Claire has over 18 years of commercial experience in the UK, including private consultancy, the public sector, and work in the third sector supporting local SMEs. She lives and works in Greater Manchester in the UK with her family.

**Carmela Cucuzzella** is the Dean of the Faculty of Environmental Design at the Université de Montréal. Prior to her current position, she was Professor in the Department of Design and Computation at Concordia University where she held the University Research Chair in Integrated Design and Sustainability for the Built Environment (www.ideas-be.ca). She was also the founding co-director of the Next Generation Cities Institute at Concordia University. She is also a member of the inter-university and interdisciplinary team of the Laboratory for the Study of Potential Architecture (LEAP). Her research work is framed within the broad domain of design studies, where she investigates questions of sustainable design for urban living. Her varied background and expertise in environmental and social life cycle analysis, green building rating systems, and design and architecture allow her to adopt a framework revolving around design's interrelated dimensions of the cognitive-instrumental, the moral-practical, and the aesthetic-expressive forms of conception and discourse. In her CoLLaboratoire research, she seeks to understand how the collaborative design and implementation of interactive art-architecture in public urban spaces can contribute to a critique, a deeper understanding, and/or the embodiment of sustainable urban, professional, community, and even human practices in the long term. In her second area of research, her interests lie predominantly in responsible design practices, with a particular interest in understanding the challenges of accommodating sustainability diagnostic or rating tools such as Life Cycle Assessment (LCA) and Leadership in Energy and Environmental Design (LEED) alongside the creative conceptual exploration that takes place during the design process. She addresses the limits of current sustainability assessment tools as a means to gain a complex understanding of the social, cultural, and environmental repercussions of design practice.

**Rana Geith** is a research and teaching assistant in the Department of Architecture at the American University in Cairo (AUC), from which she recently graduated with a Bachelor of Science in Architectural Engineering. Her research interests include environmental impact assessments, form- and space-making potential for sustainable design, environmental and ecological architecture, and spatial decision-making. Her interest in sustainability stems from her belief in the importance of architecture and design in tackling complex societal challenges. She has prior work experience in architectural design, site coordination and supervision, marketing, and communication. She also served as the elected President of the Architecture Association (AY 2020-2021) at AUC's Department of Architecture.

**Sherif Goubran** is an assistant professor of sustainable architecture in the Department of Architecture (School of Sciences and Engineering) at the American University in Cairo. He completed his Ph.D. in the Individualized (INDI) Program at Concordia University in 2021. Before that, Sherif completed a MASc in Building Engineering in 2016, focusing on energy efficiency in commercial buildings. He holds a BS in Architecture from the American University in Cairo (AUC-Egypt). Sherif's research focus includes building sustainability and sustainability assessment, sustainability in architectural design, and human approaches in design. Specifically, his work investigates the theory and practice of sustainability in the built environment, combines qualitative and quantitative methodologies, and explores the shift from incremental to transformational design. He conducts interdisciplinary research within the fields of design, architecture, building engineering, and real estate finance. He is also involved in several sustainability committees and projects on administrative levels, as well as in the broader community.

**Ilija Gubić** has worked for the United Nations based in Bangkok, Manila, Nairobi, New York, and Yangon. He has also worked for the Global Green Growth Institute based in Kigali. Ilija has conducted research at Columbia University in New York City, USA, Politecnico di Milano in Italy, and the Faculty of Architecture in the University of Belgrade in Serbia. Further, he occupied the role of visiting assistant lecturer at the School of Architecture and Built Environment at the University of Rwanda. He has published works on architecture and urban development in least developed and developing countries.

**Usha Iyer-Raniga** is a professor at the School of Property and Construction Management, RMIT University in Melbourne. Usha is co-leading the One Planet Network's Sustainable Buildings and Construction Program (SBC) as part of the United Nations' 10 Year Framework of Programs on Sustainable Consumption and Production (UN 10FYP SCP) aligned with Sustainable Development Goal 12. The work also directly impacts SDG 11.

**Deepika Jauhari** is the principal architect and researcher at DARS, an Architecture and Research Studio based in New Delhi, India. Her work deals with public and private design projects primarily related to sustainability and urban development backed by research. Her professional qualifications include interdisciplinary fields of architecture, landscape architecture, and ekistics. Previously, she was associated with the Delhi Development Authority as the Assistant Director in the Housing and Urban Projects Wing. Deepika has received several merit-based scholarships and has consistently published her research works on various national and international platforms. Her research interests include urban growth and regeneration, urban planning, sustainable development, placemaking, informal settlements, and environmentally sensitive landscapes.

**Mohammad Ahmed Hussein Khatybeh** is a Master of Architecture candidate at Istanbul Aydin University. He worked as a project supervisor at Charisma Co., leading a team in a written audio project to strengthen the database of artificial intelligence. He is currently a member of the international department team, MUSAID, forming a link between international traders and the association. He is interested in formulating and implementing equipment designs, testing and producing specifications, and researching product applications.

**David LePage** is the founder and managing partner of Buy Social Canada. He is recognized internationally as a social enterprise and social procurement thought leader, and as an effective practitioner, and public policy architect. Developing and applying innovative market-based solutions over many years to complex social issues has led to David's extensive experience in multi-sector engagements, skills development, and knowledge sharing. David blends direct community-based service delivery and influencing public policy to ensure effective and measurable outcomes. His experience and knowledge are built upon years of learning and applicable practice in roles with social enterprises, intermediary services, and membership organizations. In addition to his work at Buy Social Canada, David designed the Social Enterprise Leadership M.B.A. Specialty Program at the University of Fredericton in 2013 and has been its lead instructor since. His research on social procurement has been published in the Canadian Journal of Nonprofit and Social Economy Research, the Canadian Review of Social Policy, and the Government of Canada's Horizons Policy Research Initiative, Social Economy Issue.

**Marco Mancini** is an architect with a Ph.D. in Design "cum laude." His academic and professional activity started in 2005. His main topics are product design, innovation theory, and technology for industrial design, with a special focus on the relationship between music and design. He teaches at the University of Ferrara, the Florence Fine Arts Academy, and ISIA Design in Pescara, alongside collaborations with the University of Florence. He has presented at international conferences and was a visiting professor in "Innovation in Design" at the Nanjing University of Aeronautics and Astronautics (NUAA), Nanking, China. He is an ordinary member of the Italian Design Society (SID) and a member of the Technical Commission of ISO-UNI n°10 TC/33 "Products, Processes and System for the Building" board on behalf of the Italian National Council of Architects. As a freelancer, his areas of focus include architecture, industrial/product design, and graphics. He deals with outfitting and temporary installation for international exhibitions, such as Pitti Immagine in Florence. He holds patents on stage products for music, and on integrated systems for the safety of artifacts in emergency conditions. He is a jazz pianist, composer, and singer/songwriter enrolled in the Italian Society of Authors and Editors (SIAE).

**Emma Renaerts** is the Communications Manager at Buy Social Canada. Passionate about creating social change and supporting community-building, sustainability, equity, and reconciliation, she uses her skills as a journalist and communicator to foster dialog and raise awareness about innovative solutions to today's pressing issues. Emma highlights the possibilities of a social value marketplace both online and through community collaboration. She spearheads daily social media content creation, larger storytelling projects, and qualitative interviews, and leads event planning for Buy Social Canada. In her role, Emma also supports training and communications for the British Columbia Social Procurement Initiative. Emma has a Master of Journalism from the University of British Columbia and has had stories published in The Tyee, This Magazine, and CBC Radio Vancouver.

**Ahmed Helmy Salem** is the general director of the General Organization for Physical Planning and has a Ph.D. in regional and urban planning from Cairo University. He has taught statistical analysis of scientific research in the faculties of Engineering and Urban Planning at Cairo University, alongside his lectures in the faculties of Engineering at Zagazig University, Qatar University, the University College of the Sultanate of Oman, and the Faculty of Fine Arts in Minya University. He has published and presented more than twenty papers at international conferences in various fields of urban planning. Dr. Salem's experience also includes working as an urban planner in consultancy offices in both Egypt and the Kingdom of Saudi Arabia.

**Robert Brian Smith** graduated with a B.Sc. in 1969 and has undertaken postgraduate studies in education, engineering, and law since then. Robert holds a Ph.D. in engineering and is currently a Ph.D. candidate in law at the University of New England, Australia. He has had a long career, first as a government officer for 20 years, followed by nearly 25 years as an engineering consultant. Since April 2012, Robert has been a full-time international development consultant working in Southeast and South Asia. He has been a team leader in Bangladesh, India, Myanmar, Nepal, and Sri Lanka and has spent shorter periods on projects in Cambodia, Indonesia, the Philippines, Thailand, and Timor Leste. These postings have given him unique insights into the development needs of least developed and developing economies. One of his passions is to share his skills and knowledge with others, and he has published over 120 refereed papers in engineering, international development, and law, particularly on Thai energy law, cybercrime, and intellectual property.

**Kamani Sylva** is a senior lecturer attached to the Faculty of Engineering, The University of Peradeniya in Sri Lanka. She contributes to research in the field of sustainable development through the circular economy and social resilience. She has obtained her B.Sc. from the University of Peradeniya and holds Master's degrees in Geotechnical Engineering (AIT, Bangkok), Business Administration (PIM, Sri Lanka), and Energy Systems (Gävle, Sweden).

**Bob Tomlinson** is the founder and director of The Living Village Trust, Living Villages Ltd, and Village Makers Ltd. He has won several design, building, and development awards including Housebuilder of the Year, Best Landscape Award, and Best Environmental Development. Despite prior intentions to study architecture in the 1970s, he was persuaded by his father to get some work at an architects' practice before embarking on several years of study. The designs for modernist housing that the firm were producing were so appalling that he ran away to sea and became a professional explorer and filmmaker instead. Years later, the making of a documentary on sustainable development prompted Bob to a design and proposal for a Living Village, a collection of attractive homes surrounded by food growing and recreational gardens. The idea caught on, and after dogged determination and perseverance against a construction industry set in its ways, the 40 houses surrounded by shared gardens, orchards, and woodland known as The Wintles were built. The success of this project led to others and the setting up of a development company that worked on sites throughout the UK. In over 25 years, Bob has learned by experience how to make sustainable design work in a construction industry resistant to change.

**Thomas Walker** holds an M.B.A. and Ph.D. degree in Finance from Washington State University. Prior to his academic career, he worked for several years in the German consulting and industrial sector at such firms as Mercedes Benz, Utility Consultants International, Lahmeyer International, Telenet, and KPMG Peat Marwick. His research interests are in emerging risk management, corporate finance, venture capital, sustainability and climate change, FinTech, corporate governance, securities regulation and litigation, insider trading, and institutional ownership; he has published over eighty articles, book chapters, and edited books in these areas. He is the lead editor of nine books on sustainable financial systems, sustainable real estate, sustainable aviation, environmental policy, emerging risk management, innovations in social finance, fintech, and water risk management. Dr. Walker currently serves as the principal investigator on research grants by the Social Sciences and Humanities Research Council (SSHRC), the Autorité des marchés financiers, and the Global Risk Institute. In 2018, he founded the Emerging Risks Information Center (ERIC, https://emerging-risks.com), which conducts targeted research on environmental, technological, and societal risks that affect our world today. In 2021, he became the inaugural director for the Jacques Ménard/BMO Centre for Capital Markets at Concordia University and the Concordia University Research Chair in Emerging Risk Management (Tier 1).

# Abbreviations

| | |
|---|---|
| AI | Artificial Intelligence |
| BAU | Business As Usual |
| BDP | Benefit Driven Procurement |
| BEIS | The UK Department for Business, Energy, and Industrial Strategy |
| CAD | Computer-Aided Design |
| CBA | Community Benefit Agreement |
| CEO | Chief Executive Officer |
| CERN | European Organization for Nuclear Research |
| CIBSE | The UK Chartered Institute of Building Service Engineers |
| CLT | Community Land Trust |
| CSR | Corporate Social Responsibility |
| DfD | Design for Disassembly |
| DOLIDAR | The Department of Local Infrastructure Development and Agricultural Roads |
| EMBERS | Eastside Movement for Business and Economic Renewal Society |
| ESSA | Économie Sociale et Solidaire en Aménagement |
| EU | European |
| FAR | Floor Area Ratio |
| FRAPRU | Front d'action Populaire en Réaménagement Urbain |
| g | Solar Gain Factor |
| GAP | Güneydoğu Anadolu Projesi |
| GBMCS | Green Building Minimum Compliance System |
| GHG | Greenhouse Gas |
| GI | Green Infrastructure |
| GPS | Global Positioning System |
| GRTs | Groupes de Ressources Techniques |

| | |
|---|---|
| HUG | Home User Guide |
| IBM | The International Business Machines Corporation |
| IBRD | International Bank for Reconstruction and Development |
| IEA | International Energy Agency |
| ILO | International Labor Organization |
| IPCC | Intergovernmental Panel on Climate Change |
| ISO | International Standard Organization |
| MDG | Millennium Development Goal |
| MHCLG | Ministry of Housing, Communities, and Local Government |
| MiR | Made in Rwanda |
| MNC | Multi-National Corporation |
| $MtCO_2e$ | Metric Tons of Carbon Dioxide Equivalent |
| NASA | National Aeronautics and Space Administration |
| NBC | Nepal National Building Code |
| NDC | Nationally Determined Contribution |
| OCPM | Office de Consultation Publique de Montréal |
| PPC | Plastic Pollution Coalition |
| PPP | Public-Private-Partnership |
| PROECCO | Promoting Climate Responsive Construction Material Production and Off-farm Employment in the Great Lakes Region |
| PV | Photovoltaic Panel |
| PVL | Photovoltaic Laminate |
| RAI | Rural Access Index |
| RFID | Radio Frequency Identification |
| RFP | Request for Proposals |
| SDC | Swiss Agency for Development and Cooperation |
| SDG | Sustainable Development Goal |
| SHE | Sane, Humane, Ecological |
| SHGC | Solar Heat Gain Coefficient |
| SME | Small and Medium-Sized Enterprise |
| SNRTP | Strengthening the National Rural Transport Programme |
| SPSS | Statistical Package for the Social Sciences, a statistical software suite |
| SUP | Single Use Plastic |
| TBMM | Grand National Assembly of Turkey |
| TOKI | Housing Development Administration |
| TUIK | Turkish Statistical Institute |
| UCLA | University of California, Los Angeles |
| UK | United Kingdom |

| | |
|---|---|
| UN | United Nations |
| UNCED | 1992 UN Conference on Environment and Development |
| UNDP | United Nations Development Programme |
| UNEP | United Nations Environmental Programme |
| US/USA | United States of America |
| USGBC | U.S. Green Building Council |
| WHO | World Health Organization |
| WPR | World Population Review |
| WSUD | Water-Sensitive Urban Design |
| WWR | Window-to-Wall Ratio |

# List of Figures

**How Rwanda's Green Building Minimum Compliance System Can Help Achieve the Sustainable Development Goals**



**Determining an Adequate Population Density to Achieve Sustainable Development and Quality of Life**



**Achieving the UN SDGs Through the Integration of Social Procurement in Construction Projects**



**Relationship Between Changes in Building Culture and the Carbon Footprint from Past to Present: A Case Study from Şanlıurfa-Turkey**

### Partnership in the Built Environment for Realizing the 2030 Agenda: A Soft Systems Model Incorporating Systems Theory and the Circular Economy



### Equity in the Built Environment in Least Developed Countries: The Case of Rural Municipalities in Nepal

# List of Tables

### Relationship Between Changes in Building Culture and the Carbon Footprint from Past to Present: A Case Study from Şanlıurfa-Turkey



### Equity in the Built Environment in Least Developed Countries: The Case of Rural Municipalities in Nepal

# Toward Realizing the 2030 Agenda in the Built Environment: An Introduction

Thomas Walker, Carmela Cucuzzella, Sherif Goubran, and Rana Geith

## Introduction

The publication of the UN 2030 Agenda has set a unified framework for governments, businesses, civil society, and researchers for sustainable development action (Diaz-Sarachaga et al., 2018; United Nations, 2015; Wysokińska, 2017). Since its adoption in 2016, it was expected that the Agenda would influence politics, capital flows, and development priorities globally and that it would trickle down to direct the focus and action of many non-government actors (Bojer, 2017; Gupta & Vegelin, 2016; Jayasooria, 2016). In an area where there were many competing definitions and criteria (Bernardi et al., 2017; Doan et al., 2017), the UN 2030 Agenda and its 17 Sustainable Development Goals (SDGs) set a new milestone for sustainability in the built environment, providing a stable, expansive and precise definition (Diaz-Sarachaga et al., 2018). Additionally, the Agenda's

T. Walker (✉)
Department of Finance, Concordia University, Montreal, QC, Canada
e-mail: thomas.walker@concordia.ca

C. Cucuzzella
Faculty of Environmental Design, Université de Montréal, Montreal, QC, Canada
e-mail: carmela.cucuzzella@concordia.ca

S. Goubran · R. Geith
Department of Architecture, The American University in Cairo, New Cairo, Egypt
e-mail: sherifg@aucegypt.edu

R. Geith
e-mail: ranag@aucegypt.edu

 
T. Walker et al. (eds.), *The Role of Design, Construction, and Real Estate in Advancing the Sustainable Development Goals*, Sustainable Development Goals Series, https://doi.org/10.1007/978-3-031-28739-8_1

goals and targets were designed around what is commonly known as the 5 Ps: people, planet, prosperity, peace, and partnership—thus dedicating almost equal attention to the environmental, social, and economic pillars (Jayasooria, 2016). For the last few years, governments and development agencies have turned to the built environment and to practitioners and researchers who work in the field to accelerate the realization of the global sustainability agenda.

The increasing attention the built environment, especially urban centers and cities, has received is well justified since it intersects with many critical challenges such as population growth, urbanization, and un-sustainability (Goubran et al., 2019). Buildings and real estate have significant adverse consequences for the natural environment, as major contributors to global greenhouse gas emissions, large consumers of natural resources and materials, and key generators of waste and water effluents (World Economic Forum, 2016). Moreover, the built environment and its activities significantly overlap with the social, economic, cultural, and political dimensions of sustainability (Ehrenfeld, 2009; Intergovernmental Panel on Climate Change, 2015; James, 2014; Moore & Engstrom, 2004). Thus, tackling the built environment's sustainability challenges can alleviate significant environmental damages and provide social, economic, cultural, and health benefits (Goubran et al., 2019). However, realizing these improvements requires that actors work quickly, collaboratively, interdisciplinarily, and systematically to mobilize building activities to facilitate attaining *sustainability* in the built environment (Fisher, 2008; Fry, 2014; Intergovernmental Panel on Climate Change, 2015; Sustainable Development Solutions Network Thematic Group on Sustainable Cities, 2015).

The sustainable built environment has been studied extensively in the last decades. Many researchers have been interested in the built environment's potential to contribute to the SDGs (Alawneh et al., 2019; Goubran, 2019). Available studies have looked at how practices, processes, materials, and operations can help the realization of specific SDGs and their targets (Goubran & Cucuzzella, 2019; Omer & Noguchi, 2020). Researchers also point out that the built environment can contribute to many of the SDGs' targets by capitalizing on developmental synergies, adopting systemic approaches, and meaningfully including more stakeholders (Allen et al., 2018; Fuso Nerini et al., 2018; Guinée, 2016). Moreover, given the 2030 Agenda's call for transformative change, scholars are now exploring moving the building and real estate industry beyond the traditional environmental risk management approaches (Goubran et al., 2023).

Yet, much of the available research focuses on the ecological and environmental dimensions of sustainability, limiting its ability to bridge between the 5Ps of the 2030 Agenda (Bernardi et al., 2017; Bragança et al., 2010). Furthermore, recent publications have often followed thematic or disciplinary approaches and primarily focused on specific phases of building projects. However, to understand the contribution of building projects in attaining the SDGs, the built environment has to be understood as an enabler of sustainable development and one has to capitalize on its substantial role in shaping the daily lives of communities. To do that, projects' complete life cycles must be considered, including their design, construction, investment, management, and regulatory dimensions. Additionally, a holistic and interdisciplinary outlook has to be adopted to investigate innovations in processes, tools, and policies.

This edited book uncovers how real estate projects' design, construction, operation, and investment can accelerate the realization of the SDGs in the built environment. It examines a broad range of emerging innovations, methods, and approaches for integrating the SDGs in the life cycle of projects. It also proposes how practices and theories beyond the traditional academic fields of engineering, finance, economics, and political science can play an integral role in attaining global goals, and how interdisciplinarity when dealing with the built environment can result in critical synergies that accelerate the institutionalization of sustainable development practice. The collection highlights how new market solutions, public policies, and design practices can be mobilized to realize the technological and social infrastructures required to attain the SDGs. It offers a transition beyond the traditional "green building" standards when defining sustainable development in the built environment. Considering the global nature of the topic and its multiscale consequences, the book offers scalar relevance, including entries that discuss the local, regional, national, and supranational levels. Through this approach, the book reveals the transformative capacity of built environments and the drivers and barriers for the meaningful integration of the SDGs across the building and real estate sectors.

## Overview of Content

Following this introduction, the collection is divided into three thematic sections, each addressing different phases of the built environment life cycle. The content of each of these sections is summarized below.

This first section, *Designing More Sustainable Human Settlements,* discusses new approaches and strategies to enhance the sustainable development of cities and communities to accommodate the needs of the people and provide solutions to rising challenges to human settlements.

The section begins with the chapter "The Potential of New Methodologies, Approaches, and Artificial Intelligence Techniques in Addressing the Urgent Development Challenges of the Built Environment", which examines the relationship between current trends in urban development and the advancement of SDG 11, a goal that encourages sustainable city and urban development through the design of inclusive, safe, and resilient communities. Bob Tomlinson puts the current city planning approaches into question. He criticizes how architects and city planners, when faced with time and budget constraints, sometimes produce designs that are not meeting the principles set by SDG 11 for safe, resilient, and inclusive environments. The author explains how Computer Aided Design (CAD) and algorithm-based Artificial Intelligence (AI), if programmed with environment pattern elements, can lead to sustainable models of the built environment, generating buildings that are well-adapted to their specific contexts. The chapter further emphasizes these innovative approaches by integrating a development solution for slum areas, showcasing how SDG 11 can be achieved through design connected with local communities' culture, climate, and aspirations.

The chapter "Achieving the Agenda 2030 in the Built Environment: Role, Benefits, and Challenges in Implementing Green Infrastructure in Informal Settlements", focuses on the role of green infrastructure (GI) in achieving the SDGs, specifically discussing its relevance and applicability in the development of informal settlements. Deepika Jauhari starts by highlighting the impact of GI in offsetting some of the negative effects related to urbanization, climate change, and rapid population growth. The author then elaborates on the role of GI in advancing the SDGs by examining the targets that are relevant to SDGs 6, 7, 8, 9, 11, 13, and 15, and analyzing how they directly or indirectly contribute to the SDGs. The chapter relies on Delhi in India as its base case, and presents similar cases across Asia, Africa, and South America to demonstrate how GI can be implemented in informal settlements to accelerate attaining the SDGs.

The final chapter of this section, "Solar Shading Design and Implementation in UK Housing as a Tool for Advancing Sustainable Development", sheds light on the important role solar shading plays in achieving the SDGs. Claire Brown investigates how simple interventions such as solar shading can reduce emissions and accelerate the sustainable development of the built environment. The chapter further explores how solar shading serves as a practical solution for ensuring the resilience of exposed homes in the UK to the effects of climate change. The author further highlights the potential of this fundamental design element in elevating the impact of the residential building stock, and minimizing the risk of overheating in housing units.

The book's second section, *Built Environment Policy and Governance Innovations,* explores the SDGs' advancement in the built environment by implementing environmental policy and governance innovations that enhance sustainable practices and life cycle approaches.

The chapter "How Rwanda's Green Building Minimum Compliance System Can Help Achieve the Sustainable Development Goals", puts forth green building features that contribute to achieving the SDGs in building projects in Rwanda. The study responds to the rapid urbanization trend in Rwanda, which is expected to increase buildings' greenhouse gasses (GHG) emissions by 574% by 2050 if no corrective actions are taken. Ilija Gubic and Dheeraj Arrabothu demonstrate how recently adopted policies, such as the Green Building Minimum Compliance System (GBMCS) and Made in Rwanda (MiR) building materials helped to reduce cumulative emissions. Moreover, these policy solutions also helped excel the industry's environmental, social, economic, and health standards, enhancing social inclusion and advancing the country's progress toward the SDGs.

In his chapter "Determining an Adequate Population Density to Achieve Sustainable Development and Quality of Life", Ahmed Salem studies the relationship between adequate urban population density and the achievement of the SDGs, specifically SDGs 3, 7, and 11. The chapter shows that high-density urban settlements encompass several advantages, including walkability, enhanced social interaction, better public transit, and better availability of services for the community. However, high-density communities also have disadvantages, as they may lead to higher stress levels for their inhabitants. The author advocates the use of statistical and spatial analysis programs to calculate optimized urban density levels that balance these advantages and disadvantages and thus help to realize sustainable development in cities.

The last chapter of this section, "Achieving the UN SDGs Through the Integration of Social Procurement in Construction Projects", explores the concept of social value in public policy directives. Focusing on social procurement tactics in construction projects, the chapter encourages transitioning from traditional efficiency-based construction methods to more conscientious decision-making. David LePage and Emma Renaerts emphasize the importance of favoring social procurement in construction projects through training, social endeavors, and employment. They highlight that social procurement contributes to increased employment opportunities, training, and more sustainable supply chains, which lead to healthier and more sustainable communities. The study relies on case studies of policies and practices across Canada to demonstrate the forces driving this proposed shift and repurpose the construction industry. The chapter concludes by indicating that measuring social value helps realize the SDGs and create inclusive, equitable, and diverse communities.

The last section of the collection, *Systemic Approaches to the Built Environment Systems,* illustrates frameworks that represent systemic approaches to the built environment systems, provides assessment strategies for comparing buildings' performance, and outlines sustainable building techniques for sustainable development.

The chapter "Relationship Between Changes in Building Culture and the Carbon Footprint from Past to Present: A Case Study from Şanlıurfa-Turkey", reviews vernacular housing building samples in Turkey and compares them to contemporary construction using passive system parameters. Specifically, Mohammad Ahmad Hussein Khataybeh and Alpay Akgüc explore the impact of changes in building culture on the energy efficiency and carbon emissions of buildings in Şanlıurfa, Turkey. They propose a rating system based on climate-responsive building design strategies and utilize it to develop an algorithm for reducing the adverse effects of greenhouse gas emissions. This approach helps decrease $CO_2$ emissions based on the action plans stipulated in the UN SDGs' Goal 7 (Accessible and Clean Energy) and Goal 11 (Sustainable Cities and Communities).

The chapter "Partnership in the Built Environment for Realizing the 2030 Agenda: A Soft Systems Model Incorporating Systems Theory and the Circular Economy", highlights the critical role that partnerships between the multiple stakeholders of the built environment play in achieving sustainable cities and communities. Kamani Sylva and Usha Iyer-Raniga elaborate on how the built environment can meet the UN 2030 Agenda through a strategic vision that combines all the sustainable development goals. The chapter proposes a conceptual framework for a soft systems model that incorporates systems theory with a circular supply chain through partnerships at different levels.

Next, the chapter "A Sustainable Approach to the Planning, Organization, and Management of Big Events in the Music Entertainment Industry", offers a new perspective on how big events can be organized in a sustainable fashion. The study also shows how the COVID-19 pandemic caused the suspension of many major events, which brings about an opportunity to reopen these event areas with a dedication to the sustainable development of the environment. It calls for using new technologies and innovations that provide environmental improvements as part of event management, where the social dimension is a focal reason for public appeal. Marco Mancini calls for resetting big events post-pandemic by focusing on the SDGs and by capitalizing on positive environmental outcomes.

Finally, the chapter "Equity in the Built Environment in Least Developed Countries: The Case of Rural Municipalities in Nepal", directs attention to the development of accessibility infrastructure in rural areas, focusing on Nepal as its case study. Robert Brian Smith shifts the discussion of the built environment, which is often focused on large population centers such as cities, to smaller and often neglected rural areas. The chapter highlights the means of implementing the SDGs on a micro-level when dealing with the least developed countries and areas. The study proposes strategies to improve accessibility in the built environment for marginalized rural areas to promote the advancement of the SDGs. The chapter concludes by providing a framework that prioritizes accessibility, which can be mobilized for advancing sustained and impactful development in the least developed countries' rural centers.

## References

Alawneh, R., Ghazali, F., Ali, H., & Sadullah, A. F. (2019). A Novel framework for integrating United Nations Sustainable Development Goals into sustainable non-residential building assessment and management in Jordan. *Sustainable Cities and Society, 49*, 101612. https://doi.org/10.1016/j.scs.2019.101612

Allen, C., Metternicht, G., & Wiedmann, T. (2018). Initial progress in implementing the Sustainable Development Goals (SDGs): A review of evidence from countries. *Sustainability Science, 13*(5), 1453–1467. https://doi.org/10.1007/s11625-018-0572-3

Bernardi, E., Carlucci, S., Cornaro, C., & Bohne, R. A. (2017). An analysis of the most adopted rating systems for assessing the environmental impact of buildings. *Sustainability, 9*(7), 1226. https://doi.org/10.3390/su9071226

Bojer, M. (2017). What if we really meant it? Transformative approaches for the Sustainable Development Goals. In *Reos Partners Blog*. Reos Partners. https://reospartners.com/really-meant-transformative-approaches-sustainable-development-goals/

Bragança, L., Mateus, R., & Koukkari, H. (2010). Building Sustainability Assessment. *Sustainability, 2*(7), 2010–2023. https://doi.org/10.3390/su2072010

Diaz-Sarachaga, J. M., Jato-Espino, D., & Castro-Fresno, D. (2018). Is the Sustainable Development Goals (SDG) Index an adequate framework to measure the progress of the 2030 agenda? *Sustainable Development*. https://doi.org/10.1002/sd.1735

Doan, D. T., Ghaffarianhoseini, A. A., Naismith, N., Zhang, T., Ghaffarianhoseini, A. A., & Tookey, J. (2017). A critical comparison of green building rating systems. *Building and Environment, 123*, 243–260. https://doi.org/10.1016/j.buildenv.2017.07.007

Ehrenfeld, J. R. (2009). Sustainability by design: A subversive strategy for transforming our consumer culture. *Yale University Press*. https://doi.org/10.1111/j.1530-9290.2009.00209.x

Fisher, T. (2008). *Architectural design and ethics: Tools for survival*. Elsevier/Architectural Press.

Fry, T. (2014). *Design futuring sustainability, ethics and new practice.* Bloomsbury Academic.

Fuso Nerini, F., Tomei, J., To, L. S., Bisaga, I., Parikh, P., Black, M., Borrion, A., Spataru, C., Castán Broto, V., Anandarajah, G., Milligan, B., & Mulugetta, Y. (2018). Mapping synergies and trade-offs between energy and the Sustainable Development Goals. *Nature Energy, 3*, 10–15. https://doi.org/10.1038/s41560-017-0036-5

Goubran, S. (2019). On the role of construction in achieving the SDGs. *Journal of Sustainability Research, 1*(2). https://doi.org/10.20900/jsr20190020

Goubran, S., & Cucuzzella, C. (2019). Integrating the Sustainable Development Goals in building projects. *Journal of Sustainability Research, 1*(e190010), 1–43. https://doi.org/10.20900/jsr20190010

Goubran, S., Masson, T., & Caycedo, M. (2019). Evolutions in sustainability and sustainable real estate. In T. Walker, C. Krosinsky, L. N. Hasan, & S. D. Kibsey (Eds.), *Sustainable real estate* (pp. 11–31). Palgrave Macmillan. https://doi.org/10.1007/978-3-319-94565-1_3

Goubran, S., Walker, T., Cucuzzella, C., & Schwartz, T. (2023). Green building standards and the United Nations' Sustainable Development Goals. *Journal of Environmental Management, 326*, 116552. https://doi.org/10.1016/j.jenvman.2022.116552

Guinée, J. (2016). Life cycle sustainability assessment: What is it and what are its challenges? In R. Clift & A. Druckman (Eds.), *Taking stock of industrial ecology* (pp. 45–68). Springer Open. https://doi.org/10.1007/978-3-319-20571-7

Gupta, J., & Vegelin, C. (2016). Sustainable Development Goals and inclusive development. *International Environmental Agreements: Politics, Law and Economics, 16*(3), 433–448. https://doi.org/10.1007/s10784-016-9323-z

Intergovernmental Panel on Climate Change. (2015). Climate change 2014: Mitigation of climate change. In O. Edenhofer, R. Pichs-Madruga, Y. Sokona, J. C. Minx, E. F. Head, S. Kadner, K. Seyboth, A. A. Team, I. Baum, S. Brunner, P. Eickemeier, B. Kriemann, J. Savolainen, S. Schlömer, C. von Stechow, & T. Z. Senior (Eds.), *Climate change 2014 mitigation of climate change. Part of the working group III contribution to the fifth assessment report of the intergovernmental panel on climate change.* Cambridge University Press. https://doi.org/10.1017/CBO9781107415416

James, P. (2014). Urban sustainability in theory and practice. *Routledge.* https://doi.org/10.4324/9781315765747

Jayasooria, D. (2016). Sustainable Development Goals and social work: Opportunities and challenges for social work practice in Malaysia. *Journal of Human Rights and Social Work, 1*(1), 19–29. https://doi.org/10.1007/s41134-016-0007-y

Moore, S. A., & Engstrom, N. (2004). The social construction of "Green building" codes: Competing models by industry, government and NGOs. In *Sustainable architectures: Critical explorations of green building practice in Europe and North America* (pp. 51–70). Spon Press Taylor & Francis Group. https://doi.org/10.4324/9780203412800

Omer, M. A. B., & Noguchi, T. (2020). A conceptual framework for understanding the contribution of building materials in the achievement of Sustainable Development Goals (SDGs). *Sustainable Cities and Society, 52*(January), 101869. https://doi.org/10.1016/j.scs.2019.101869

Sustainable Development Solutions Network Thematic Group on Sustainable Cities. (2015). The urban opportunity: Enabling transformative and sustainable development. *Sustainable Development Solutions Network*. https://sustainabledevelopment.un.org/content/documents/2579Final-052013-SDSN-TG09-The-Urban-Opportunity.pdf

United Nations. (2015). Transforming our world: The 2030 agenda for sustainable development. *United Nations*. https://sdgs.un.org/2030agenda

World Economic Forum. (2016, January). Environmental sustainability principles for the real estate industry. *World Economic Forum*.

Wysokińska, Z. (2017). Millenium Development Goals/UN and Sustainable Development Goals/UN as instruments for realising sustainable development concept in the global economy. *Comparative Economic Research, 20*(1), 101–118. https://doi.org/10.1515/cer-2017-0006

# Designing More Sustainable Human Settlements

# The Potential of New Methodologies, Approaches, and Artificial Intelligence Techniques in Addressing the Urgent Development Challenges of the Built Environment

Bob Tomlinson

## Introduction

Sustainable Development Goals (SDGs) have always been the guiding objectives behind settlements. For over 6000 years we have been designing and building cities based on political, religious, military or commercial ideologies, with a concern for longevity. When industrialization accelerated urbanization, settlements incorporated elements of social welfare as concerns arose about the health and well-being of residents. More recently, growing awareness of the environmental damage caused by human activities has meant that sustainability has been formally introduced as a desired building element.

Sustainability is, however, a very difficult thing to ensure by design alone. Despite aspirations for longevity, settlements from ancient Mesopotamia onwards have not survived in their original form. They have decayed and disappeared, or they have evolved and adapted to changes in climate, culture and economic conditions (Kostof, 1991). This ability to respond to change is the key to longevity and sustainability. Cities that have managed to evolve share a unique pattern. The pattern of streets, dwellings, meeting places, markets, workplaces, schools, parks, and local amenities form an integral part of a local community that has responded to cultural, economic and climatic changes (West, 2017). Because of this, these communities have a greater sense of ownership, inclusivity, safety and resilience

B. Tomlinson (✉)
Village Makers—Design and Build Practitioners, Devon, UK
e-mail: bob.tomlinson@village-makers.com

 
T. Walker et al. (eds.), *The Role of Design, Construction, and Real Estate in Advancing the Sustainable Development Goals*, Sustainable Development Goals Series, https://doi.org/10.1007/978-3-031-28739-8_2

(Girardet, 2004). Any new settlement design which hopes to genuinely deliver on the United Nations Sustainable Development Goals (SDGs) must understand the needs and aspirations of the local community and the ways in which these are expressed in their neighborhoods.

The damage caused by the pandemic's lockdowns and isolation of communities has shown just how important human connectivity and integration are to the economic and social well-being of settlements. The design of new places to live and urban regeneration projects need to take these human needs into account. This increases the ability of these new settlements to adapt and ensures their long-term sustainability.

## A New Form of Urban Design

Landry (2006) states that "City making is an art, not a formula". Rigid formal urban design, with grid-based layouts, zones, and focus points, does a poor job of providing the right conditions for communities to evolve sustainably as it cannot adapt easily to changing conditions. There are valid reasons for this fact, which will be discussed later (Howard-Kunstler, 1993). However, a combination of Global Positioning Systems (GPS), Computer Aided Design (CAD), Artificial Intelligence (AI) and 3D printing can facilitate a new form of urban design that works with the natural evolution of a settlement and allows long term sustainability aligned with the delivery of the Agenda 2030 sustainable development goals.

## The Problem

Conventional, top-down urban design development and regeneration has tended toward relatively simplistic modernist grid and block layouts that suit the control and administration of policy, or commercial development (Krier, 1998). Emulating the complex patterns and small details of an evolved settlement is a very time-consuming process with which urban designers and architects are unfamiliar. Even if there is a commitment to invest in more sustainable design at the planning stage, the "real-world" problems of procurement, regulatory compliance, and complying financing at the build stage, usually rule out the construction of neighborhoods based on these intricate patterns of human development.

## Using Technology with the Design Solution

To address these problems at the urban design stage, existing technology needs to be used in combination with a community consultation process. Such an approach will be a direct response to community patterns of use and interaction with the site. This process delivers an urban form that is in tune with the behaviors, cultures, politics, and aspirations of the local community as well as taking into account the physical topography, microclimate and constraints of the site.

## Using Computer Aided Design Together with 3D Printing for the Build Solution

The build stage of a project could then be handed over to the residents. Such an approach would replicate the way in which traditional organic settlements evolve. This may work if the "ownership" of the land parcels can be allocated together with a strict design code in order to control the outcomes, as may be the case in more affluent communities. In less affluent communities, however, this would only result in a new kind of slum where homes are built from whatever materials are readily available at the least cost. To ensure the build quality of new homes in poorer areas is maintained, a contractor is usually employed to complete the construction. In this case, cost efficiencies usually lead to a repetition of identical dwellings on a homogenous grid layout that pays little heed to local culture, climate or topography. To avoid this, artificial intelligence and computer-aided design linked to on-site delivery methods, particularly 3D printing, can evolve a series of home types. These are then modified into unique units that fit the natural topography and layout derived from a community consultation process.

## Planned Settlements

There is an immense body of work which addresses the design of urban forms, and its different objectives as society, politics and commerce have changed over time. In ancient times, cities were laid out in reverence to deities or colonizing authorities (Kostof, 1991). In the eighteenth century, wealthy landowners demolished existing villages and replaced them with picturesque replacements to improve the view from their dwelling (Darley, 1978). The nineteenth century saw philanthropists and shrewd industrialists improving the living conditions of the working class by designing more hygienic layouts (Wilson, 2020). In the twentieth century, modernists with visions of concrete utopias, aided by politicians and entrepreneurs interested in cheaper construction methods, brought about a new way of building cities. More recently, biophilic urbanism has promoted the concept of nature-based environments in the urban form (Stevens-Curl, 2018).

All these approaches have been generally well-intended, presenting practical solutions to urban problems, but decisions are often made on behalf of local populations without consulting them directly. Communities have thus had little to no say in the design of their neighborhoods (Jabareen, 2006).

## The Cosmic City Form

To put this information into context, let us imagine a design brief for a new place to live that delivers the objectives of The United Nations Agenda 2030 Goal 11. Key elements of the brief are to make cities and human settlements inclusive, safe, resilient, and sustainable. Given this brief, intelligent and creative minds would probably start thinking about the exciting prospects of what Kevin Lynch (1984) termed the *cosmic* model of city design. The *cosmic* model comes from the idea that the settlement plan comes from a pure design source, spatial diagrams of hierarchical structures, sometimes derived from symbolism. In any event, the plan is likely to have a monumental axis focussing on civic areas surrounded by a strictly defined and regular grid, often mirrored around the center. Some urbanists argue that many ancient settlements originated in this way as a form shaped by symbolism rather than practical design, citing examples such as ancient Athens, Mexico City, and various Imperial Cities. Over the ages, cities like these have been laid out according to religious or military diagrams, inspired by modern standards of urban planning like Brasília's bird or airplane layout (Epstein, 1973), or shamelessly commercial like Fernando Romero's Bitcoin City (Tegel, 2022).

The problem with these kinds of city layouts is that settlements designed around formal shapes and patterns tend to fracture and adapt outside the formal plan (Stevens-Curl, 2018). As Spiros Kostof (1991) points out, all settlements change over time into a form that suits the day-to-day activities of the inhabitants unless strict and permanent control is exerted over the land ownership and management of that land is continued. Inevitably, this ownership breaks down as empires die, control of policies lapse, or social and economic conditions change, eventually leading to a more organic form. However, this may take many generations having to live in uncomfortable or inappropriate surroundings.

## The Machine City Form

The second of Kevin Lynch's city forms, a practical or *machine* city, has a better track record in terms of its survival, but often falls short as far as inclusiveness, beauty, or attractiveness as a place to live. A practical settlement has functionality as the driving force behind the design, zoned for efficiency. Being able to measure lots, or parcels of land, easily and mark them out as a grid makes real estate sales and administration much more straightforward. Since the industrial revolution, colonial settlements, speculative company towns, new settlements and urban extensions have found that this is the most efficient method of development. While efficient at providing living and working space, the scale and layout of these places tend to suit the local authorities, commercial real estate, and the wealthy, leaving ordinary inhabitants with a soulless living environment. The sprawling nature of these places is also heavily dependent on fossil fuels for commuting between zones, an aspect which compromised their functionality and sustainability. Daniel Hertz (2016) has shown that in the USA, the "costs of sprawl" alone is close to $1 trillion and is rising rapidly alongside the cost of fuel. This cost is leading to the abandonment of the outer edge of sprawl as citizens spend greater proportions of their income just getting to and from work. Sprawl also has several negative impacts on commuters' families, and home lives as time spent commuting takes time away from families. Less time for exercise and leisure also undermines health and well-being and the environment suffers from pollution from numerous vehicles driving to and from workplaces.

One significant thing to learn from a place created on a grid pattern is the success of this system in allocating land ownership and control. Urbanists often extol the advantages of the gridiron layout which also suits authoritarian governments, military units, and real estate purveyors. Urbanist Paul Knight (2017) shows how the grid makes efficient use of building land by rationalizing the land used up in obtaining access to the grid blocks. Knight also points out that although land use efficiency is maximized by using super blocks, it also minimizes land devoted to rights-of-way. However, this fact minimizes the street frontage, those important social spaces where human interaction takes place to form a community.

We might thus conclude that there is an optimum grid size for resilience and sustainability that increases street frontage, but the *machine* model of practicality tends toward the creation of unpleasant places to live and the undermining of humanness. This is especially in the provision of housing in deprived areas, despite the good intentions of the architects and policymakers. One of the most famous examples of this type of failure is the Pruitt-Igoe Housing Project, demolished on live TV after only 20 years of occupation (Jenks, 1977). Out of this project came Oscar Newman's (1972) work on the importance of defensible spaces which explained why neighborhoods adjacent to Pruitt-Igoe didn't suffer the same fate because of small scale and greater detail in the relationships of spaces and buildings in contrast to the big blocks of Pruitt-Igoe. There is a very great deal more to designing a good place to live, which will be both resilient and sustainable, than focussing on the practical elements of zoning and grid systems. Newman's work is an example of how successful design takes a contextual approach and deals with not only the physical but the cultural, climatic, historical, emotional, political, and financial aspects that affect a project to make it successful.

Grids and zones are also vulnerable to changes in social and economic conditions. As the large employment generating businesses in the zones fail or relocate to more advantageous locations, the surrounding grids are abandoned. An example of this phenomenon is Detroit, perhaps the archetype of a *machine* city. As the motor industry failed, the surrounding grids became depopulated and a network of desire lines, commonly described by Robert Macfarlane (2013) as "paths & tracks made over time by the wishes & feet of walkers, especially those paths that run contrary to design or planning" (Furman, 2018), began to show themselves in defiance of the original grid layout. A striking example of this is shown by satellite photographs of abandoned blocks in Detroit when desire paths are emerging as a pattern that distorts the grid layout.

## The Organic City Form

The transition of parts of a *machine* city into a more natural form as it responds to changing conditions is described in Kevin Lynch's third model, the *organic* or biological city. This model is seen as a living thing, an organism that seeks to maintain a balanced state in response to change. Unlike the other two models, whereby the *cosmic* model requires constant and rigorous land control, and the *machine* model becomes obsolete as times change, the *organic* model can adapt and be sustainable over time. The difficulty with the *organic* model is that by nature it is either anarchical, with the potential of resulting in a slum type of development, or is the planned picturesque of garden suburbs whose resilience to change, in truth, is questionable.

The big advantages of the *cosmic* and *machine* models for settlements, are that, given sufficient support from political and economic potentates, it is possible to deliver a built solution. The downside is that this is very much a top-down model of development, undermining inclusivity, and the ability of a place to have that complex mix of human and environmental interaction that determine well-being as set out by Christopher Alexander (1965). Without the key ingredient attractiveness to potential residents, who must also have a sense of belonging and control over their immediate surroundings, places will not be sustainable. At best, they will be modified and altered by the residents, at worst they are demolished, resulting in great environmental and social harm. We can also see that layouts that try to run over the natural desire lines also fail in the long term when social, climatic, and economic factors force land control to slip from authoritarian ownership. One of many examples of the inevitability of desire lines to override a grid in the long term is the way in which the medieval agrarian settlement pattern has been superimposed over a Roman grid in the German town of Trier (Kostof, 1991).

## Desire Lines as a Basis for Design

Returning to our Design Brief, it would seem sensible to learn the lessons from *cosmic* and *machine* cities and understand that, sooner or later, a natural shape determined by desire lines will dominate. Jane Jacobs was a strong advocate of configuring settlements around desire lines, "There is no logic that can be superimposed on the city; people make it, and it is to them ... that we must fit our plans" (Mehaffy, 2008). Accepting that desire lines will eventually override geometrically designed solutions is a way of ensuring long-term resilience and sustainability in a new layout. Danish urban planner Jan Gehl (2010) describes following desire lines as "listening to a place" and he focuses on the natural interactions between people and spaces. In other examples, architect Rem Koolhaas (1997) allowed the paths made by the footfall of students over time to inform his plan for the Illinois Institute of Technology. Andrew Furman (2018) notes that desire lines point to "the endless human desire to have choice."

Ultimately, Desire Lines also deliver the most energy-efficient and eco-friendly route between two points. In the past, engineers thought that straight-line paths were the most sustainable, but we now realize that gradients, ground conditions, and focus points will create a desire line that takes the most energy-efficient and interesting route around the soft ground, rocks, or toward points that we don't fully understand but find attractive (Colville-Andersen, 2018). In construction too, creating a desire line lead design approach results in the minimum of cut and fill, which is an expensive and environmentally damaging exercise.

Desire lines have other environmental and personal health benefits. Avoiding motor vehicle use is key, but to encourage the use of footpaths and cycleways they need to be a safe, attractive, useful, and efficient route. Desire lines serve as a natural basis for these paths. Jeff Speck (2003) points out the fact that "it turns out that trading all of your incandescent light bulbs for energy efficient savers conserves as much carbon per year as living in a walkable neighbourhood does each week. Transportation energy use consistently tops household energy use, in some cases by 2.4 to 1. As a result, the greenest home (with Prius) in sprawl still loses out to the least green home in a walkable neighbourhood."

None of this information is new. Most inhabitants of the pre-industrial world would be familiar with a living place developed around desire lines. As the machine age dawned, visionaries like Camillo Sitte (1889) identified what works in cities and connects patterns in architecture and planning with patterns of social life. Even the great grid advocate Le Corbusier (1929) had to admit that the pack donkey could be the true architect of the cities of Europe. "Man walks in a straight line because he has a goal and knows where his is going […] the pack-donkey meanders along, meditates a little in his scatter-brained and distracted fashion, he zigzags to avoid the larger stones, or to ease the climb, or to gain a little shade; he takes the line of least resistance. […] The Pack-Donkey's way is responsible for the plan of every continental city." (Le Corbusier: The City of Tomorrow and its Planning 1929). And, as we have seen from Detroit, over time that may also be true for more recent cities as well (Fig 1).

## Introducing the Design Team to the Concept

An important factor in the design of a new settlement as a dynamic and inclusive assessment of desire lines, topography, climate, and culture, rather than a series of lines on a drawing, is to get the design team to "connect" with the site. I do this by getting the design team on site and then asking them to think about where they would like to make their home, considering the topography, sunlight paths, prevailing winds and so forth. For the first time, the designers connect their thoughts on a new layout with the actual, physical elements of the earth and existing community, rather than seeing the site as pixels on a computer screen or lines on a drawing. Using GPS (NASA, 2019), we can then pinpoint on a plan where the design team thinks they would like to live, and then follow in their footsteps to plot the desire lines which connect these to the access points. With this exercise, the beginnings of an organic layout for an undeveloped site can be

**Fig. 1** A desire line showing the natural path across a field with an "unexplainable" kink. Photograph by Carole Salmon

explored as part of the formal design for infrastructure and access (Figs. 2, 3, 4, 5, 6 and 7).

## Using GPS in Practical Application

When the design team is comfortable with an organic, responsive approach to the urban layout, the next stage is to include the existing or proposed community and test the designers' assumptions. Building on the methodology developed by organizations like Planning for Real (2018), where a dynamic, 3D model-based

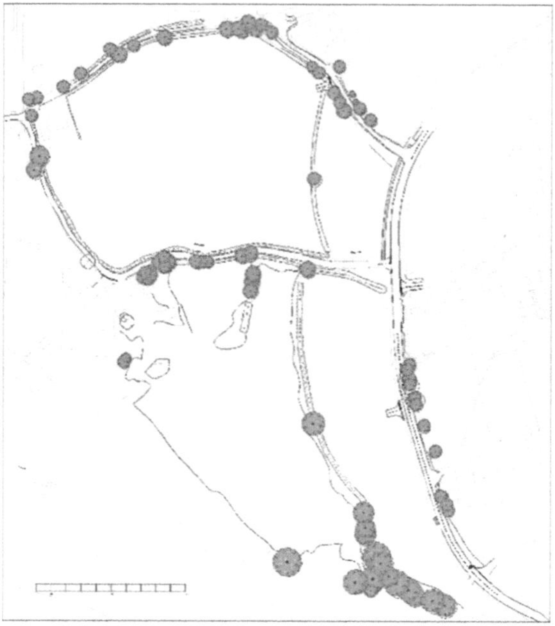

**Fig. 2** An example of the Tamar Village site without existing development. Village Makers. Bob Tomlinson

method is used to involve communities and allow them to have a real say about what happens in their neighborhoods, we can update their ideas with GPS technology and computer-generated imagery to integrate the community activities and desires into a dynamic design process. The UN-Habitat report (2004), acknowledged that "urban planning can become a tool for inclusion and slum prevention if it is designed to tap informality as a development force and guide it towards the making of better cities." In line with this advice, the significant element here is that by using the model and relating it to the project, the process is inclusive. This approach contrasts with a conventional design process that relies on architects' drawings and layout plans which are difficult to read for most non-professional attendees. For fear of appearing ignorant, many people are thus deterred from commenting or expressing an opinion about the design proposals.

It is vital that the local community and local authority get involved in the project and help supply the data via GPS tracking which will shape the new

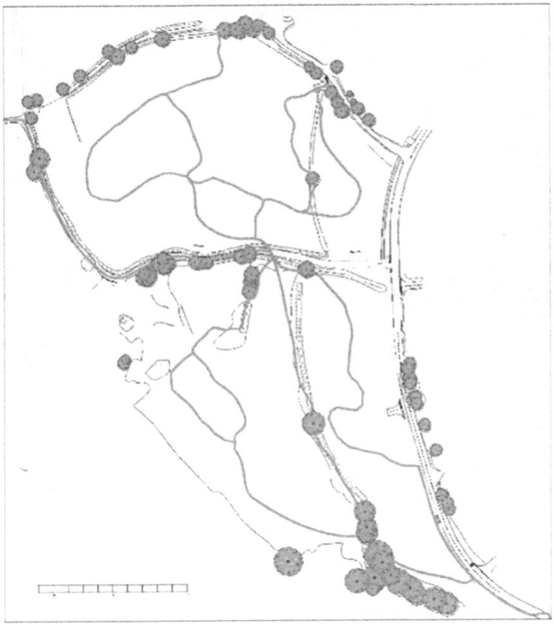

**Fig. 3** Walking the site and tracking the route creates desire lines. Village Makers. Bob Tomlinson

neighborhoods rather than be passive observers of a process. The dynamic element is to ask them to contribute by allowing the anonymous recording of their travel through the area to build up a pattern of movement that begins to form the basis of a new layout. The success and popularity of applications like Strava (2022) and mapmywalk (2022) show that, with the proper presentation, this process can be widely accepted (Davies, 2014). A highly successful example of community involvement, where residents were trained to collect data, is in Mukuru in Nairobi where every latrine, water tap, and electricity pole in the settlement was mapped (Muungano Alliance, 2021). Having a simple scoring system like a heart or cross could also be incorporated into the GPS tracking. This will also connect residents to the process and encourage them to identify areas that they like, or dislike, within the development site.

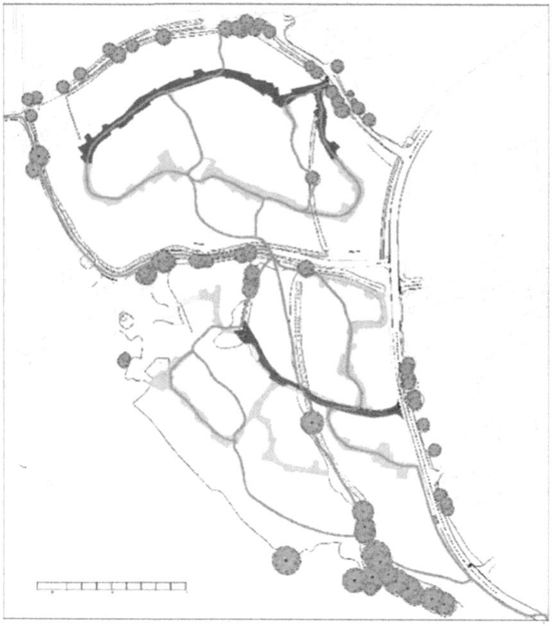

**Fig. 4** Vehicle, cycle and pedestrian routes are designed around the desire lines. Village Makers. Bob Tomlinson

## Established Use of GPS Tracking

GPS has been used to track movement in health and welfare monitoring situations, to map things like geographical access to public health resources and transport services, or to monitor health in a particular environment. Such uses of GPS include recording human patterns of movement related to health behavior, for example tracking elderly people suffering from dementia or Alzheimer's disease to prevent wandering or analyzing sedentary behavior (Apte, 2019). With the recent pandemic, track and trace systems have been accepted by most of the population to deter behaviors that increase the transmission of Covid. Increasingly, transport designers and engineers are using GPS data to find those routes that people prefer to use between their homes to work, or school, shopping, or leisure at different times of day and compare them to those that have been designed. For example, Bike Citizens (2022) has introduced a GPS data analysis tool for traffic

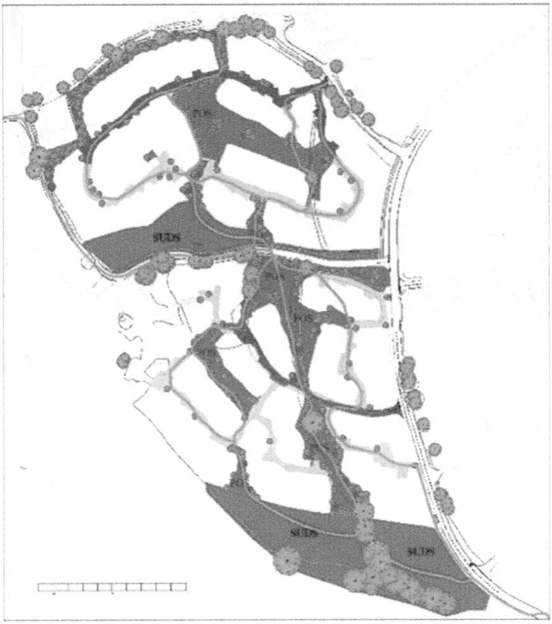

**Fig. 5** Shared open space, new landscape planting and sustainable urban drainage is added. Village Makers. Bob Tomlinson

planning by providing a visual portrayal of which routes cyclists are taking as well as an overview of bicycle traffic behavior.

Of course, this process relies on GPS and the use of mobile "phone data but as 91.54% of people in the world are now cell phone owners, this should be possible even in the poorest of areas (Statista, 2022)." As with the health monitoring to record patterns of movement, GPS devices can be given to those without cell "phones and most people will also be generally familiar with GPS as a primary navigation aid, or as a Google Maps link from a multitude of applications needing to give a location (Google Maps, 2022)."This use of GPS technology to observe desire lines through a new site or existing neighborhood, brings an active and "readable" dynamic to the community consultation process. This development must be done in the spirit of SDG No. 17, Partnerships for the Goals, which will only work if there is trust and integration from the local people. The challenge is to make sure that all participants understand that the ethical issues are covered,

**Fig. 6** New buildings are plotted using the network of access routes. Village Makers. Bob Tomlinson

that the data will be securely anonymous, and that unnecessary data will be destroyed (Apte, 2019).

For the design team, it must also be accepted that the data tracking movement may shed light on certain issues which will need to be dealt with discretion. For example, patterns of movement might unveil a pattern of numerous visits to a place, shop, or service the popularity of which its users would prefer to go unnoticed. Ensuring the anonymity of participants is thus paramount, but aspects of a neighborhood or community that had hitherto been hidden, may nonetheless be revealed. It will be important, therefore, to accept that even potentially concerning behaviors and movements shown by the GPS tracking are present within the community and need to be taken into consideration when a neighborhood is designed. Whatever the aspirations of the designers and authorities, patterns of behaviors are likely to persist unless these are addressed in the design. For example, if the GPS tracking shows that antisocial behavior occurs in a particular area,

**Fig. 7** Houses at The Wintles fronting a desire line. Village Makers. Bob Tomlinson

it would be a good idea to analyze why this happens and seek out design solutions in the same way that Newman (1972) did at Pruitt-Igoe rather than ignore issues.

It may well be that existing networks of routes through neighborhoods result from desire lines over many years. We may therefore be able to use GPS to track movements through neighborhoods and allow urban designers to assess the effectiveness of the existing network. This possibility is supported by UN-Habitat's strategy brought forward by Claudio Acioly (2020) which has "streets as the natural conduits that connect slums spatially and physically with the city and treats streets not only as a physical entity for mobility and accessibility, through which water and sewerage pipes, power lines, and drainage systems are laid, but also as the common good and the public domain where social, cultural and economic activities are articulated, reinforced and facilitated."

These existing routes must not be ignored; they should form the basis of the replacement development. The relationship of the private areas of the home to the public street or square is an important part of the sustainability of cities, especially in slum areas. Streets are where social connections are made which lead to work and entrepreneurial opportunities. Ignoring these relationships, and assuming that slums can be replaced with high-rise blocks on a sterile grid, has caused many slum replacement projects to fail (UN-HABITAT, 2010). If the project is looking at the regeneration of a site, superimposing the pathways of movement through an existing layout will reveal preferred routes through an existing settlement and suggest how a replacement layout should respond to those routes to best serve the community. Assuming complete demolition of a non-functioning neighborhood, desire lines can still be the basis for the replacement lattice of pathways for the new development.

## Studying What Works for the Community

Research for a project must also include a detailed study of local places, streets, neighborhoods, and public spaces that have proven themselves to be sustainable over several generations. This process is one of recording and considering why some places are preferred, and others are not, looking particularly at the relationship of the building form to the public space. This study should be done as close to the site as possible and make up the first stage of a slum clearance or improvement project. If this is not possible in the locality because no good examples exist, or the area is uninhabited, the study should progressively move outwards from the site in order to find those places that show that they have been cared for. Considering the aspirational elements of a local community when it comes to desirable living spaces is also vital.

In large areas that are devoid of habitation or have been blighted by inappropriate development in the past, it may be necessary to travel some distance to find an example of a truly sustainable place, but it is important that such a place be within the same cultural and climatic zone. These are the places that have accommodated cultures, social and economic changes, and micro-climates in ways that are resilient, and they will provide a great deal of useful information as to what accounts for such resilience. Thus, such a process will allow designers to identify which places have been loved and cared for over many generations, and which ones have been neglected.

Generic examples of the type of patterning resulting from this mix of human and environmental interaction that determine human well-being are set out by Christopher Alexander (1978). While A Pattern Language sets out the method with examples of successful archetypical patterns, it is vital to use this method to generate a unique set of patterns that are finely tuned to the geographic location for the project because local cultures and climate will modify the process. The more accurate this can be, the more likely the buildings and neighborhoods will be a success and be embraced by the new occupants and be sustainable in the long term. In our own experience at The Wintles we used A Pattern Language to inform the layout design followed by the study of much-loved local vernacular to achieve the desired result. After almost 20 years of occupancy residents still enthuse about the neighborhood (Wintles, 2022).

## Community Consultation Together with Desire Lines

Having amassed a baseline of information to provide a background platform for the site, it will now be important for designers to introduce a thorough community consultation to reveal a great deal more about the opportunities and constraints of the site. It is at this stage that community participation is crucial in helping us to understand the design challenges, which may include elements normally obscured in a more formal top-down process. These elements may be obvious, like the physical elements of topography, access and infrastructure, but a responsive consultation, starting with desire points and lines brings out much more information. These climatic and cultural elements will need to be addressed in the design if the project is to be properly inclusive as well as sustainable. The exercise of introducing the proposed project as an organic evolution of the natural form of the site to accommodate the needs and aspirations of the users is very different from the usual top-down development. This process allows a much more human, connected, and fundamental approach to the project (Malone, 2019).

Having incorporated the information from the community consultations into the design, designers will be able to tailor their design to the needs and aspirations of those people, which means that they will feel more connected to their new home. They are more likely to have a sense of ownership over it, as well as take care and positive long-term relationship with the new place. It may be that this community consultation and feedback will also highlight where the problems are by identifying undesirable patterns of use which need to be addressed in the design. For example, things like SDG No. 6, Clean Water and Sanitation can be considered within the context of real use, in that patterns of use will demonstrate how effective and appropriate design solutions can be implemented. These will have a greater chance of success than a solution merely devised on a drawing board.

## The Skeletal Layout

What we now have should be a network of paths through a neighborhood that are finely tuned to the ways in which they are used by the community. These can be plotted onto the site plan and form the basis of the emerging layout. In the same way that Corbusier (1929) described the natural formation of settlements growing up along his derided pack donkey tracks. "Along these tracks, houses are 'planted,' and eventually these houses are enclosed by city walls and gates. Five centuries later another larger enclosure is built, and five centuries later still a third yet greater." (The City of Tomorrow and its Planning Page. Corbusier, 1929) we can begin to think about how the homes, businesses, facilities and amenities of the new neighborhoods can be delivered onto the site as buildings. Our problem is that we don't have several centuries to achieve this evolved, organic growth. We need to build quickly, viably, and responsibly to create the new sustainable neighborhood.

## Using Artificial Intelligence

Up until this point, the design team's challenge has been a largely manageable process, covering the usual site surveys, community consultations, constraints plans, viability studies and introducing the designed layout conventionally, all-be-it derived from the unconventional more organic process. It is quite common for this phase of a development to describe many good things for the local community and environment which are then diluted and often discarded as the practical forces of procurement and delivery take hold. In a conventional project, this usually occurs because of the expense and time-consuming elements of thorough community consultation, and because detailed design response is often seen as an expensive luxury and one which produces designs that can never be built because the proposed projects are too expensive.

This is no longer the case. Recent advances in Computer Aided Design and algorithm-based Artificial Intelligence mean that design systems can be programmed with the successful built environment's pattern elements as described by the design code and used to generate appropriate building designs. Instead of trying to design the whole project as one piece of work, a small number of buildings are 3D modeled in detail using the local patterning derived from the observation and community consultation. Thereafter, in the same way that settlements evolve in response to population growth, topography, economic activity, culture and climate stimulus, the models can be iteratively adapted by the software to suit the development site and recreate that sustainable form of organic place-making that has proven itself to be so resilient in the past.

Software designer Bruno Postle (2013) defines this process as Informal Urbanism. He explains that "The approach described here is to use Pattern Languages as a fitness criteria to guide an iterative design optimisation." To complement this, a Form Language as described by the physicist Nikos Salingaros (2013) can be used as the basis for information used by the AI to determine iterative changes to the designs. Michael Mehaffy (2019) also notes how this challenged the notion of "design" as a progressive expression of schematic intentions and argued for a conception of design as a stepwise, non-linear evolution in response to a series of contextual urban factors.

Bruno Postle (2013) describes how these then work with CAD systems. This is a process for domestic buildings that is "physically buildable, sufficiently flexible, and suitable for evolution through mutation and crossover," he explains. Postle continues, "the idea is to take clues from the same processes that form buildings in historical cities and towns: i.e., rather than developing the design in a single step, start with a simple building and modify it iteratively using small incremental improvements."

It is quite usual for commercial housebuilders to have a range of house types and modify these on a plot-by-plot basis for wealthy open market developments to improve placemaking and marketability. In poor social housing, this is rarely the case. Repetition of type without modification to suit the plot environment, prevailing wind, sunlight, views, or cultural elements all comes at a cost that is usually deemed to be excessive.

Using AI software, however, designers can modify a small number of house-types through the CAD process which responds to the patterns generated by the layout and the Form Language as part of the program to create neighborhoods or streets that are responsive to topography, connecting routes, micro-climates, access to daylight, views, privacy and so forth with as much flexibility and detail as the system programmers will allow (Postle, 2013).

With access to the 3D model and the ability to take off dimensions, local builders, self-builders and contractors should be able to construct the new neighborhood efficiently using traditional construction methods. There are advantages to this in that it provides employment, and where the supply conditions allow, a building specification requiring the use of local sustainable materials will reduce the environmental impact (Figs. 8 and 9).

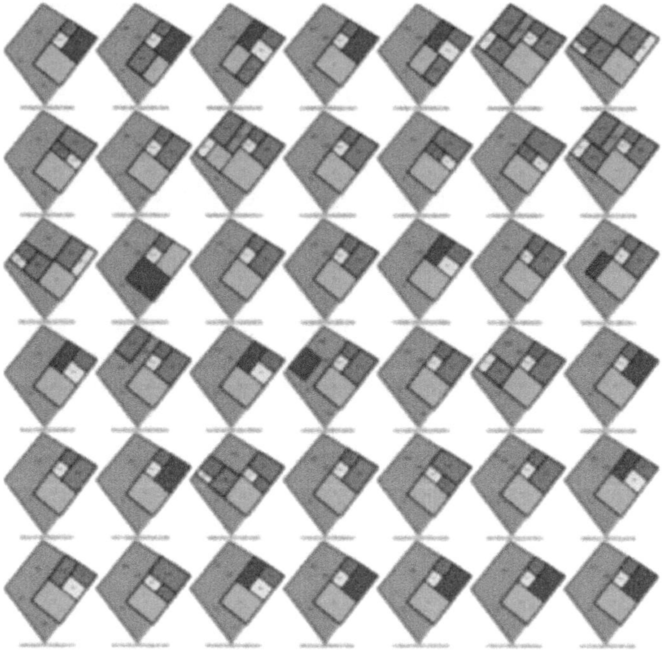

**Fig. 8** Population during the evolution of a single-story house after 48 generations. Bruno Postle

## 3D Printed Buildings

In areas where labour availability or costs are an issue, an emerging development can link the 3D CAD modeling directly to a 3D printer capable of printing out a complete building or its components. The advantages of this process are that the 3D printing machine can be programmed with a unique task for each individual plot while taking into account all the many variations inherent to these sites, in order to make the best use of that plot without having to issue unique sets of drawings each time. Compared to traditional methods, where labor cost is an issue, there are great time savings from using a 3D printer to do the main part of the construction work. For example, Apis Cor (2022) has been able to deliver a 40 square meter home in 24 hours with an overall build cost of 25% of conventional build using a mobile 3D printer.

**Fig. 9** Rendered view of a generated cluster of seven buildings. Bruno Postle

The other significant advantages of 3D printing are the reduction of waste, and the integration of energy efficiencies and sustainable materials into the process. The design of the buildings can be based on the life cycles of the materials to improve the environmental sustainability of building materials (WASP, 2022). The buildings can be a unique shape, made to be the best fit for a particular part of the site (though the relationship with adjoining buildings will begin to determine a more uniform pattern) but the natural topography and existing elements of the site can be accommodated.

Environmentally friendly construction processes can also be utilized by programming the printer to use local raw materials with low embodied energy, waste, and minimal water usage. On-site injuries are thus minimized, and hazards from dust and handling heavy materials are reduced.

## Conclusion

Just like any complex system, or living organism, the ability to adapt to change is crucial to the survival of human settlements. When we take the long-term view, thousands of years, we can see that cities that have grown up organically and changed with times tend to be more resilient. This is because of the intense network of social connections and micro scale of buildings that enable the population to adapt to climatic, cultural and economic changes. Industrial age cities with their sprawling infrastructure and heavy dependence on fossil fuel are not a good model for future settlements but our design and build systems are generally locked into this paradigm.

However, by using GPS tracking with the population the "organic" patterns and nature of a desired settlement can be relatively quickly and cheaply determined. This gives a structure for a settlement layout that will be energy efficient as well as one preferred by the residents. This will be a response to a proposed development that has been rooted in the behaviors of the people who will live there. It will be a place that facilitates social interaction and therefore inclusiveness. It will be a safe place because of the human scale, and it will be resilient through the possibility of easily adapting to change.

Following the design work, the building of the unique homes, streets, squares and focal points that go into any desirable community can be achieved more cost-effectively by using computer-aided design. A series of building types are developed and cost and then a series of algorithms are introduced so that the types can be adapted to individual plots determined by the street layout. The algorithms can be programmed to take into account things like topography, microclimates, cultural requirements, overlooking, opportunities for adaption and expansion, and so forth. In this way, the buildings respond much more accurately, and efficiently, to a site and its inhabitants. The scale of the build process means that local materials and labor are more likely to be used or 3D printed housing can be utilized to improve environmental benefits. Every home can be unique, enhancing a sense of place, ownership and responsibility. This builds community resilience and identity and is far more effective at delivering the objectives of Agenda 2030s SDG number 11, inclusivity, safety, resilience, and sustainability.

## References

Acioly, C. (2020). *Street-led citywide slum upgrading* https://doi.org/10.1007/978-981-13-7307-7_2

Alexander, C. (1965). *A city is not a tree.* Sustasis Press.

Alexander, C. (1978). *A pattern language.* Oxford University Press USA.

Apis Cor. (2022). https://apis-cor.com/contact-us/

Apte, A. (2019). *Ethical considerations in the use of GPS-based movement tracking in health research.* https://jogh.org/documents/issue201901/jogh-09-010323.pdf

Bike Citizens. (2022). https://www.bikecitizens.net/

Colville-Andersen, M. (2018). https://colvilleandersen.medium.com/my-desire-line-development-dream-5ba364e94980
Corbusier, L. (1929). *The city of tomorrow and its planning*. Dover Publications.
Darley, G. (1978). *Villages of vision*. Granada Publishing.
Davies, A. (2014). *Strava's cycling app is helping cities build better bike lanes*. https://www.wired.com/2014/06/strava-sells-cycling-data/
Epstein, D. (1973). *Brasilia: Plan and reality*. University of California Press.
Foster, A., & Newell, J. (2019, September). *Detroit's lines of desire: Footpaths and vacant land in the motor city*. Landscape and Urban Planning.
Furman, A. (2018). Desire paths: The illicit trails that defy urban planners. https://smartgrowth.org/desire-paths-the-illicit-trails-that-defy-urban-planners/
Girardet, H. (2004). *Cities people planet*. John Wiley.
Gehl, J. (2010). *Cities for people*. Island Press.
Google Maps. (2022). https://maps.google.com
Hertz, D. (2016). *How much does urban sprawl cost American Commuters*. https://www.bloomberg.com/news/articles/2016-06-08/city-observatory-calculates-cost-of-urban-sprawl-for-commuters-in-u-s-metros
Howard-Kunstler, J. (1993). *The geography of nowhere*. Touchstone.
Jabareen, A. Y. (2006). Sustainable urban forms: Their typologies, models, and concepts. *Journal of Planning Education and Research, 26*, 38. https://doi.org/10.1177/0739456X05285119
Jenks, C. (1977). *The language of post-modern architecture*. Rizzoli.
Knight, P. (2017). *Defending the American Grid*. www.smartcitiesdive.com
Koolhaas, R. (1997). *The McCormick Tribune Campus Centre*. https://www.architonic.com/en/project/oma-the-mccormick-tribune-campus-center/5100219
Kostof, S. (1991). *The city shaped: Urban patterns and meanings through history*.Thames and Hudson.
Krier, L. (1998). *Architecture choice or fate*. Andreas Papadakis.
Landry, C. (2006). *The art of city making*. Cromwell Press.
Lynch, K. (1984). *Good city form*. The MIT Press.
Malone, L. (2019). *Desire lines: A guide to community participation in designing places*. RIBA Publishing.
Mapmywalk. (2022). https://www.mapmywalk.com/
Macfarlane, R. (2013.) *The old ways: A journey on foot*. Penguin.
Mehaffy, M. (2008). Generative methods in urban design: A progress assessment. *Journal of Urbanism International Research on Placemaking and Urban Sustainability, 7*(2).
Mehaffy, M. (2019). *Cities alive: Jane Jacobs, Christopher Alexander, and the roots of the new urban renaissance*. Mijnbestseller.
Muungano Alliance. (2021). https://www.muungano.net/publicationslibrary/v
NASA. (2019). *What is GPS*. https://www.nasa.gov/directorates/heo/scan/communications/policy/GPS.html
Newman, O. (1972). *Defensible space*. Macmillan.
Planning For Real. (2018). https://www.involve.org.uk/resources/methods/planning-real
Postle, B. (2013). *An adaptive approach to domestic design*. https://journalofbiourbanism.org/2014/01/01/jbu-volume-ii-12-2013/
Salingaros, N. (2013). *Unified architectural theory*. Vajra Books.

Sitte, C. (1889). *The art of building cities: City building according to its artistic fundamentals.* Hyperion Press.
Speck, J. (2003). *Walkable city: How downtown can save America, one step at a time.* North Point Press.
Statista. (2022). https://www.statista.com/statistics/330695/number-of-smartphone-users-worldwide/
Stevens-Curl, J. (2018). *Making dystopia.* Oxford University Press.
Strava. (2022). https://www.strava.com/
Tegel, S. (2022, May 10). *Behold the Bitcoin City: El Salvador's 'millennial dictator' reveals plans for futuristic metropolis.* Telegraph Newspaper.
The Sustainable Development Goals Report. (2020). https://sdgs.un.org/publications/sustainable-development-goals-report-2020-24686
UN-Habitat report. (2004). https://unhabitat.org/un-habitat-self-assessment-report-2004
UN-HABITAT. (2010). *Reinventing urban planning for sustainable cities.* https://unhabitat.org/sites/default/files/download-manager-files/1404131088wpdm_Planning%20Sustainable%20Cities%20UN-HABITAT%20Practices%20and%20Perspectives.pdf
WASP. (2022). https://www.3dwasp.com/en/3d-printed-house-gaia/
West, G. (2014). https://www3.weforum.org/docs/WEF_GAC_GlobalAgendaOutlook_2014.pdf, p. 23.
West, G. (2017). *Scale: The universal laws of life, growth, and death in organisms, cities, and companies.* Penguin Press.
Wilson, B. (2020). *Metropolis.* Penguin Random House.
Wintles. (2022). http://www.thelivingvillagetrust.com/the-wintles/

# Achieving the Agenda 2030 in the Built Environment: Role, Benefits, and Challenges in Implementing Green Infrastructure in Informal Settlements

Deepika Jauhari

This chapter discusses the various definitions of green infrastructure, highlighting its significance and exploring its applicability within urban built environments. Further, the chapter analyzes the role of GI in achieving the Agenda 2030 by evaluating the targets associated with the 17 SDGs (mainly SDGs 6, 7, 8, 9, 11, 13, and 15), establishing a direct or indirect contribution of GI to the SDGs. The chapter takes Delhi (India) as its case example while drawing parallels from other regions (Asia, Africa, and South America) to understand the applicability of GI in informal settlements. Thus, the chapter can be used as a reference point to understand the interactions between GI and the Agenda 2030, alongside the integration of GI in informal settlements to achieve the Sustainable Development Goals.

## Introduction

More than half (~54%) of the population of the globe currently lives in cities and urban areas. These regions are the "center of sustainability crisis: economic, ecological, political and cultural" (Filho et al., 2019, p. 72). Though each city has

D. Jauhari (✉)
Indian Society of Landscape Architects (ISOLA), New Delhi, India
e-mail: deepj23@live.in

DARS, New Delhi, India

 
T. Walker et al. (eds.), *The Role of Design, Construction, and Real Estate in Advancing the Sustainable Development Goals*, Sustainable Development Goals Series, https://doi.org/10.1007/978-3-031-28739-8_3

a different character, they all ultimately confront the same global issues: rapid urbanization, climate change, inadequate supply of basic services, social inequality and the lack of essential public goods, scarcity of resources, environmental pollution and degradation, and risks to human health (Armour et al., 2014).

Urbanization has increased the pressure on ecosystems, especially those within city boundaries. Recent urban growth, primarily driven by urbanization in the Global South (Nagendra et al., 2018), is a significant impediment to the sustainable development of cities. The result is shrinking green spaces, habitat fragmentation, pollution or contamination of natural environments, and a decline in biodiversity (Maes et al., 2019). "The urban challenge lies in meeting infrastructure and service delivery needs efficiently and effectively, balancing environmental considerations and sustaining economic and inclusive growth" (Chand Sandhu et al., 2016, p. x) while maintaining social equity.

There is growing recognition of the value of nature and urban open spaces within our built environment. The kind of ecosystem services they have to offer to urban areas (Armour et al., 2014) are integral to improving the sustainability and resilience of the built environment. Linked to this is the potential for open spaces to act as "Green Infrastructure (GI) to complement gray infrastructure spatially and at the same time counterbalance some of the adverse effects associated with gray infrastructure" (Cameron & Blanuša, 2016, p. 337). Green infrastructure refers to spaces in and around urban areas that offer a range of multi-functional benefits for a broad diversity of demographic groups in the form of ecosystem services. Green infrastructure offsets the adverse effects of rapid urbanization, unrestrained population growth, and climate change, and contributes to sustainable development and urban resilience. Though the concept of open spaces as green infrastructure is relatively new, many regions worldwide are incorporating green infrastructure into the urban built environment in order to achieve the multiple environmental, economic, and social benefits to advance the United Nations' 2030 Agenda for sustainable development.

While there is an abundance of literature to explain green infrastructure in the broader context, the contribution of green infrastructure toward the 2030 Agenda for sustainable development and sustainable urban areas is not widely known. This chapter will focus on cities and built urban environments in order to explore the interaction between Green Infrastructure (GI) and the 2030 Agenda for sustainable development.

It is essential to analyze the concept of GI with respect to regions, to explore the components of GI, to evaluate the benefits of GI, and to understand and highlight the contribution of GI to the advancement of the UN's Agenda 2030 for sustainable development. The first part of the chapter sets the stage for understanding the concept of GI in terms of commonly used definitions, as well as the applicability, benefits, scale, and components or elements of green infrastructure.

The second part of the chapter establishes the relationship between GI and sustainable development by analyzing the role of GI in achieving and advancing Agenda 2030. Furthermore, it establishes the correlation between GI and the various goals and targets of Agenda 2030 for sustainable development.

The last part of the chapter elaborates on the need for introducing GI into informal settlements, which form an essential and intrinsic part of the urban areas, especially the Global South. It focuses on the need for green infrastructure in informal settlements to achieve sustainable development of informal settlements while drawing parallels from various regions with significant informal settlement presence identifying the various limitations and challenges posed in implementing green infrastructure.

## Concept and Definition of Green Infrastructure

The concept of green infrastructure originated in the United States in the 1990s, emphasizing the "life support" functions provided by the natural environment (Kramer, 2014). GI was adopted in the UK, Australia, and recently in the European Union and with limited research and application in the Global South (Shackleton et al., 2021). This limited deployment leaves crucial gaps in the understanding of how GI and similar concepts can be applied in diverse contexts (Diep et al., 2019, p. 558). Consequently, the interpretation and definition of GI vary from region to region depending on the degree of planning and the dominant global issues affecting the region. The most popular definitions are listed in Table 1.

### Role, Benefits, and Trade-Offs

"Unlike gray infrastructure which serves a single function [green infrastructure] enhances and synergizes benefits provided by nature and achieves multiple goals with a single investment" (Kramer, 2014, p. 11). Table 2 lists the various role and benefits of GI in the urban built environment.

## Applicability, Scale, and Limitations

The design of green infrastructure should be appropriate to its context and scale depending on the underlying goals: "improving the quality of existing spaces or creating new assets which contribute to the wider network; largescale investment

**Table 1** Definitions of GI across different regions

| Region | Author | Definition |
|---|---|---|
| USA | Conservation Fund and the USDA Forest Service Green Infrastructure Work Group (1999)<br>Source: Benedict and McMahon (2001) | Green Infrastructure is our natural life support system—an interconnected network of waterways, wetlands, woodlands, wildlife habitats, and other natural areas; greenways, parks and other conservation lands; working farms, ranches and forests; and wilderness and other open spaces that support native species, maintain natural ecological processes, sustain air and water resources and contribute to the health and quality of life for America's communities and people |
| UK | National Planning Policy Framework (Ann 2, p. 69) | A network of multi-functional green space, both new and existing, both rural and urban, which supports the natural and ecological processes and is integral to the health and quality of life of sustainable communities |
| UK | Town and Country Planning Association (2004) | The subregional network of protected sites, nature reserves, green spaces, and greenway linkages |
| UK | Natural England (2009) | GI is a strategically planned and delivered network comprising the broadest range of high-quality green spaces and other environmental features. It should be designed and managed as a multi-functional resource capable of delivering ecological services and quality of life benefits required by the communities it serves and needed to underpin sustainability |

(continued)

**Table 1** (continued)

| Region | Author | Definition |
|---|---|---|
| Australia | Australian Institute of Landscape Architects<br>Green Infrastructure Report (2012) | Green Infrastructure describes the network of natural landscape assets which underpin the economic, socio-cultural and environmental functionality of our cities and towns-i.e., the green spaces and water systems which intersperse, connect and provide vital life support for humans and other species within our urban environments |
| Europe | European Commission (2013, p. 2).<br>Green Infrastructure (GI): Enhancing Europe's Natural Capital | A strategically planned network of natural and semi-natural areas with other environmental features designed and managed to deliver a wide range of ecosystem services. It incorporates green spaces (or blue if aquatic ecosystems are concerned) and other physical features in terrestrial (including coastal) and marine areas. On land, GI is present in rural and urban settings |

or smaller incremental projects; temporary, phased, or permanent interventions" (Armour et al., 2014, p. 63). The elements of a GI require working at various scales for planning and designing, namely: state- and regional-scale, city- and community-scale, and project-scale, yet "they must normally have a certain critical mass and connectivity potential to be able to contribute effectively (European Commission. Directorate General for the Environment, 2014, p. 9)."

Research has shown that developing and implementing such an ambitious project is challenging (Davies et al., 2006). Some constraints in implementing GI are related to differing socio-cultural values, different traditions and perceptions, lack of capacity, governance issues, urban planning concerns, social inequality, ecosystem disservices, spatial trade-offs, and conflicts (du Toit et al., 2018).

**Table 2** Role and benefits of GI

| Role | Benefits |
|---|---|
| Human health and well-being | GI provides vital places for recreation and physical exercise, reduces stress and improves mental health |
| Community liveability | GI can make the community liveable and cohesive by making it aesthetically pleasing, reducing air and water pollution and enhancing thermal comfort |
| Economic prosperity | Evidence shows that well-designed GI enhances the economic attractiveness of commercial precincts, increases residential property values, and creates improved tourism and economic regeneration opportunities |
| Climate modification | GI assists in reducing temperatures in cities through shading, evapotranspiration and wind speed modification while also providing protection during extreme weather events, reducing water runoff and flooding, and improving air quality |
| Water management and hydrology | GI provides efficient alternatives to conventional engineering infrastructure in the process of integrated water cycle management and Water Sensitive Urban Design (WSUD) |
| Urban ecology and biodiversity | GI supports biodiversity by creating or conserving habitat patches linked by corridors, thereby reducing habitat fragmentation and linking the different ecological assets in green networks, reducing air pollution |
| Food Production | Agriculture and productive agricultural land is a form of GI that can deliver a wide range of human health and well-being benefits that go well beyond providing a secure and healthy food supply. Community gardens are one specific type of urban agriculture that provide a unique range of social cohesion and community-building benefits. In a broader context, urban agriculture is part of 'complementary' or 'alternative' food networks that promote more sustainable practices in food production and distribution |

*Source* Compiled from M. Ely and Pitman (2014); Center for Neighborhood Technology (2010)

**Table 3** Typology and elements of GI

| Continuum | Interface | Scale | Typology of GI | GI elements |
|---|---|---|---|---|
| ↑ | ↑ | ↑ | Natural green | National parks and nature reserves, wetlands and coastal margins |
| Green | Rural | Regional | Agricultural Land | Arable land, agroforestry |
| | | | Greenways | River and creek corridors, cycleways, and routes along major transport (road, rail and tram) corridors |
| Grey-Green | Peri-Urban | City | Utility areas | Quarries, airports, and large institutional and manufacturing sites |
| | | | Sports/recreational facilities | Golf courses, school and other institutional playing fields, and other major parks. |
| | | | Squares and Plazas | Public and private courtyards and forecourts |
| | | | Streets | Residential and other |
| Grey | Urban | Project | Private/Semiprivate Gardens, | Shared (communal) spaces around apartment buildings, backyards, balconies, and community (productive) gardens |
| ↓ | ↓ | | Grey-green Solutions | Green roofs and vertical gardens, green walls, rain gardens, and bio-swales |
| | | | Engineered Solutions | Pervious paving, infiltration trench, and retention/detention ponds |

## Components of Green Infrastructure

Various networked and multi-functional green spaces constitute green infrastructure within the urban fabric and beyond. While the typology of GI varies from natural green to engineered solutions, so do the scale and gray-green continuum, as depicted in Table 3. A well-connected, well-integrated network of one or many elements can perform multiple functions and thus accrue multiple benefits in one or more environmental, economic, or social spheres contributing to sustainable development.

## Trade-Offs

According to Choi et al. (2021), the trade-offs between benefits and disservices of green infrastructure can be categorized into four aspects: (1) urban heat island

and heat stress; (2) air quality and carbon emissions; (3) water-related problems (water quality, scarcity, and groundwater recharge); and (4) health/well-being and economic disadvantages. Some examples of these include seasonal temperature reductions in winter leading to increased heating demands because of existing vegetation; reduced ventilation causing increased local temperature and thermal discomfort when trees are densely combined with shrubs/grasses (heat-trapping effects); and decreased renewable energy potential of rooftop solar panels. In addition, in the case of informal settlements, introducing green infrastructure to improve living conditions can lead to rising housing costs and property values resulting in gentrification and displacement of vulnerable residents, which require "supporting 'antigentrification' policies such as rent stabilization, housing trusts, and local employment quotas" (Hansen et al., 2017, p. 19). Thus, although the co-benefits of green infrastructure are substantially more than the trade-offs, implementing green infrastructure requires case-specific analysis.

## Transforming Our World: The 2030 Agenda for Sustainable Development

In 1987, the report "Our Common Future" produced by the Brundtland Commission defined sustainable development as development that meets the needs of the present without compromising the ability of future generations to meet their own needs (Filho et al., 2019).

As Chand Sandhu et al. (2016) elaborate, the concept of sustainability has evolved over the years. The 1990s emphasized the negative impacts of urban development and the importance of better management/protection of ecological assets. The 2000s linked environmental sustainability with economic growth, advocating "green growth" and further combining social equity with environmental sustainability and economic competitiveness. Since 2010, urban sustainability has advanced to focus on climate change resilience with the ability of cities to expand and redevelop in a way that minimizes the impact of natural hazards and climate change risks (Chand Sandhu et al., 2016). Also, 2015 became a landmark as the United Nations adopted the 2030 Agenda for Sustainable Development, replacing the Millennium Development Goals (MDGs).

The 2030 Agenda features 17 Sustainable Development Goals (SDGs) and 169 targets that set global sustainable development objectives that are integrated, indivisible, global, and universally applicable. The 2030 Agenda has five Ps as its key themes: people, planet, prosperity, peace, and partnerships. These balance

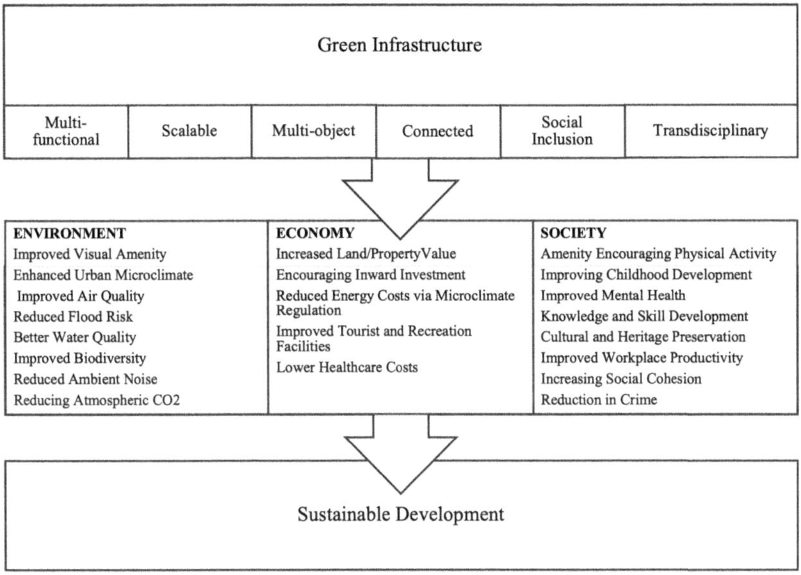


**Fig. 1** The benefits of GI in terms of environmental, economic, and social goals and how they contribute to the principles of sustainable development (*Source* Author with inputs from Armour et al. (2014); European Commission. Directorate General for the Environment [2014]; Hansen & Pauleit [2014])

the three dimensions of sustainable development: economic, social, and environmental, and with the overriding aim "to leave no one behind." The new Goals and targets came into effect on 1 January 2016 and are set to guide the decisions we make over the next 15 years (United Nations, 2015). They are comprehensive and ambitious, seeking to build on the Millennium Development Goals and accomplish what the earlier goals were not able to achieve (Fig. 1).

## Green Infrastructure and the 2030 Agenda for Sustainable Development

Cities and urban areas are required to follow the path of Sustainable development and align with SDGs in order to survive and flourish. Therefore, the role of green infrastructure in addressing the challenges of the twenty-first century cannot be overstated (Armour et al., 2014). Many authors have directly related green infrastructure as contributing to sustainability and sustainable development (see, e.g., Adegun, 2017; Benedict et al., 2012; Mell, 2009; O. Adegun, 2019) or have elaborated on the contribution of green infrastructure to the environment, economy, and society (see, e.g., Benedict & McMahon, 2001; Kramer, 2014; Mell, 2009), considered to be the pillars of sustainable development.

As per Ahern (2007), to achieve sustainability in urban landscapes, infrastructure must be seen as a means to improve and contribute to sustainability rather than just a mechanism for damage control. Green infrastructure can be seen as a key delivery mechanism for multifunctionality, contributing to the environment, the economy, and society and supporting sustainable development goals.

While there is growing global awareness of the social, environmental, and economic benefits nature can deliver in creating liveable cities (Armour et al., 2014), in order for green infrastructure to become embedded in the planning and development process, it is imperative to understand the potential of green infrastructure to contribute to achieving the 2030 Agenda for sustainable development.

## Understanding GI and SDG Interaction

### Goal 2. End Hunger, Achieve Food Security and Improved Nutrition, and Promote Sustainable Agriculture

According to Titz and Chiotha (2019), the most beneficial feature of green infrastructure has proven to be urban agriculture. Food processing plants, storage areas, and transportation facilities comprise the food infrastructure required to achieve food security (access, availability, utilization, and stability). Rooftops, balconies, public parks, road edges, and underused transportation infrastructure can be utilized for vertical farming techniques and other urban agricultural systems to help address the impending crisis in world food production, becoming integral elements in urban environments (Armour et al., 2014). Thus, the built environment can contribute to securing food supplies and preserving them.

### Goal 6. Ensure Availability and Sustainable Management of Water and Sanitation for All

Green infrastructure can foster transitions toward more sustainable and inclusive water management approaches (Diep et al., 2019). For example, green roofs can influence the urban water management cycle. GI elements like bioswales can also potentially remove pollutants in water, help build water capacity and augment supply.

## Goal 7. Ensure Access to Affordable, Reliable, Sustainable, and Modern Energy for All

Green infrastructure can serve as a passive system, reducing house heating and cooling energy demands. Moreover, as Armour et al. (2014) point out, studies have shown that the careful deployment of green infrastructure in the city can cut energy and resource costs drastically and GI can act as carbon sinks to mitigate the risks of climate change (p. 96).

## Goal 8. Promote Sustained, Inclusive, and Sustainable Economic Growth, Full, and Productive Employment, and Decent Work for All

Green infrastructure projects have the potential to create jobs that contribute to the management, design, planning, installation, maintenance, monitoring, and inspection of GI (Barnes et al., 2021). "Well-planned green space has also been shown to increase property values and decrease the costs of public infrastructure and public services" (Benedict & McMahon, p. 13). As noted earlier, however, increased property values may result in gentrification. Therefore, it is essential to continuously evaluate the site-specific impact of GI to counterbalance such unintended consequences.

## Goal 9. Build Resilient Infrastructure, Promote Inclusive and Sustainable Industrialization and Foster Innovation

Economic growth relies on building resilient infrastructure, sustainable industrialization and technological progress. The nature-based character of GI makes it reliable, sustainable, and resilient. Green infrastructure is often described as more capable than conventional "gray" infrastructure in achieving social, environmental and economic objectives (Armour et al., 2014). For example, forests and greenways are resilient and offer resilience in terms of their ecosystem service. Further, issues of urban growth and challenges in implementing green infrastructure foster innovation in GI technologies (research, prototyping, and application) (Mossin, 2018) in the form of engineered solutions and innovative materials (e.g., vertical farming, permeable paving, green concrete).

## Goal 11. Make Cities and Human Settlements Inclusive, Safe, Resilient, and Sustainable

Incorporating green infrastructure elements in "architecture, design and planning contribute in multiple ways to make cities and settlements inclusive, safe, robust, resilient and environmentally sustainable" (Mossin, 2018, p. 50).

## Goal 13. Take Urgent Action to Combat Climate Change and Its Impacts

"Many scholars and governmental organizations agree that urban green spaces, conceived as multi-functional green infrastructure, play an essential role in urban climate regulation and are accordingly helping to delineate urban climate resulting from climate change (Vásquez et al., 2019). "A green infrastructure-led design approach can be employed to weave nature into the city to provide vital carbon sinks and effective mitigation against risks such as flooding, heatwaves, and drought" (Armour et al., 2014, p. 60).

## Goal 15. Protect, Restore, and Promote Sustainable Use of Terrestrial Ecosystems, Sustainably Manage Forests, Combat Desertification, and Halt and Reverse Land Degradation and Halt Biodiversity Loss

Benedict and McMahon (2012) suggest that establishing a network of diverse green spaces might contribute to biodiversity protection, especially within urban expansion processes. "Green infrastructure protects forests, meadows, and wetlands—regarded as hubs for ecosystem services—and promotes the preservation and development of a series of less natural green spaces such as squares, parks, cemeteries, and sports facilities, which promote the existence and survival of wild flora and fauna in urban areas" (Vásquez et al., 2019, p. 335). Green infrastructure can contribute to reducing the ecological footprint of cities (Titz & Chiotha, 2019). It can potentially combat land degradation by introducing sustainable forestry, landscaped green spaces with local flora, and water management elements (like rain gardens and bio-retention ponds).

## GI and SDG Target Interaction Matrix

As identified through this study, out of the 169 targets of the 17 SDGs, green infrastructure can potentially directly contribute to 17 targets (10%) for realizing SDGs. An indirect correlation between 44 targets (26%) was observed, while almost 64% of targets had no interaction with green infrastructure. As per the study, green infrastructure can potentially contribute to advancing all 17 SDGs. If implemented appropriately, GI can make a direct contribution to the realization of SDGs 2, 6, 7, 8, 9, 11, and 15 (Table 4). Green infrastructure proved to connect most with SDG 11 of Sustainable Cities and Communities, as it has a tremendous potential to help achieve this goal (Table 5). The multi-functional, multi-object, and transdisciplinary character of green infrastructure connects with every SDG, either directly or indirectly.

## Green Infrastructure and Informal Settlements

Informal settlements are unplanned, spontaneous, unregulated settlements that do not conform to planning regulations (Ishtiyaq & Kumar, 2010; Tsenkova et al., 2009). Such settlements accommodate a substantial part of the urban population, especially in the Global South.

Globally, informal settlements are growing faster than any other form of urban development. Presently more than 1 billion people or about one in four urban dwellers live in slums or informal settlements, with 85 percent living in three regions—Central and Southern Asia (359 million), Eastern and South-Eastern Asia (306 million), and sub-Saharan Africa (230 million) (The Sustainable Development Goals Report 2022, 2022).

The reasons behind the formation of informal settlements are rapid urbanization; ineffective planning; a lack of affordable housing options for low-income households; dysfunctional urban, land, and housing policies; a dearth of housing finance; and poverty.

Unplanned settlements take the form of “squatter,” “slum,” and other “informal” settlements. A principal characteristic is that these settlements can barely sustain themselves. According to Dovey and King (2011): “a squatter lacks land

**Table 4** Matrix showing interactions between the SDG targets and GI

| SDGs | No Poverty | Zero Hunger | Good Health and Well Being | Quality Education | Gender Equality | Clean Water and Sanitation | Affordable and Clean Energy | Decent Work and Economic Growth | Industry, Innovation and Infrastructure | Reduced Inequalities | Sustainable Cities and Communities | Sustainable Consumption and Production | Climate Action | Life Below Water | Life on Land | Peace, Justice and Strong Institutions | Partnership of Goals |
|---|---|---|---|---|---|---|---|---|---|---|---|---|---|---|---|---|---|
| | **1** | **2** | **3** | **4** | **5** | **6** | **7** | **8** | **9** | **10** | **11** | **12** | **13** | **14** | **15** | **16** | **17** |
| Targets | 1.1 | 2.1 | 3.1 | 4.1 | 5.1 | 6.1 | 7.1 | 8.1 | 9.1 | 10.1 | 11.1 | 12.1 | 13.1 | 14.1 | 15.1 | 16.1 | 17.1 |
| | 1.2 | 2.2 | 3.2 | 4.2 | 5.2 | 6.2 | 7.2 | 8.2 | 9.2 | 10.2 | 11.2 | 12.2 | 13.2 | 14.2 | 15.2 | 16.2 | 17.2 |
| | 1.3 | 2.3 | 3.3 | 4.3 | 5.3 | 6.3 | 7.3 | 8.3 | 9.3 | 10.3 | 11.3 | 12.3 | 13.3 | 14.3 | 15.3 | 16.3 | 17.3 |
| | 1.4 | 2.4 | | 4.4 | 5.4 | 6.4 | 7. a | 8.4 | 9.4 | 10.4 | 11.4 | 12.4 | 13. a | 14.4 | 15.4 | 16.4 | 17.4 |
| | 1.5 | 2.5 | 3.5 | 4.5 | 5.5 | 6.5 | 7. b | 8.5 | 9.5 | 10.5 | 11.5 | 12.5 | 13. b | 14.5 | 15.5 | 16.5 | 17.5 |
| | 1. a | 2. a | 3.6 | 4.6 | 5.6 | 6.6 | | 8.6 | 9. a | 10.6 | 11.6 | 12.6 | | 14.6 | 15.6 | 16.6 | 17.6 |
| | 1. b | 2. b | 3.7 | 4.7 | 5. a | 6. a | | 8.7 | 9. b | 10.7 | 11.7 | 12.7 | | 14.7 | 15.7 | 16.7 | 17.7 |
| | | 2. c | 3.8 | 4. a | 5. b | 6. b | | 8.8 | 9. c | 10. a | 11. a | 12.8 | | 14 a | 15.8 | 16.8 | 17.8 |
| | | | 3.9 | 4. b | 5. c | | | 8.9 | | 10. b | 11. b | 12. a | | 14 b | 15.9 | 16.9 | 17.9 |
| | | | 3. a | 4. c | | | | 8.10 | | 10. c | 11. c | 12. b | | 14 c | 15 a | 16.10 | 17.10 |
| | | | 3. b | | | | | 8. a | | | | 12. c | | | 15 b | 16 a | 17.11 |
| | | | 3. c | | | | | 8. b | | | | | | | 15 c | 16 b | 17.12 |
| | | | 3.d | | | | | | | | | | | | | | 17.13 |
| | | | | | | | | | | | | | | | | | 17.14 |
| | | | | | | | | | | | | | | | | | 17.15 |
| | | | | | | | | | | | | | | | | | 17.16 |
| | | | | | | | | | | | | | | | | | 17.17 |
| | | | | | | | | | | | | | | | | | 17.18 |
| | | | | | | | | | | | | | | | | | 17.19 |
| Interaction % | 28 | 50 | 23 | 10 | 12 | 50 | 80 | 50 | 50 | 10 | 100 | 27 | 80 | 20 | 58 | 08 | 21 |
| Direct Target % | 14 | 25 | 15 | 0 | 0 | 0 | 0 | 0 | 25 | 0 | 50 | 9 | 20 | 0 | 25 | 0 | 0 |
| Indirect Target % | 14 | 25 | 8 | 10 | 12 | 50 | 80 | 50 | 25 | 10 | 50 | 18 | 60 | 20 | 33 | 08 | 21 |
| Overall Correlation | **1** | **2** | **3** | **4** | **5** | **6** | **7** | **8** | **9** | **10** | **11** | **12** | **13** | **14** | **15** | **16** | **17** |
| Sample References | Titz and Chiotha, 2019; | Armour et al., 2014; Douglas, 2018; Titz and Chiotha, 2019; | Armour et al., 2014; Schröter et al., 2016 | Demuzere et al., 2014; M. Ely and Pitman, 2014; Mossin, 2018 | Mossin, 2018 | Armour et al., 2014; Diep et al., 2019; Loftness and Haase, 2013 | Armour et al., 2014; Loftness and Haase, 2013 | Barnes et al., 202[illegible] Benedict and McMahon, 2001 | Mossin, 2018 | M. Ely & Pitman, 2014; Mossin, 2018 | Armour et al., 2014, M. Ely and Pitman, 2014; Mossin, 2018; Center for Neighborhood Technology, 2010 | European Commission. Directorate General for the Environment., 2014; Armour et al., 2014 | Armour et al., 2014, Choi et al., 2021; Vásquez et al., 2019 | Barnes et al., p. 3 | Armour et al., 2014; Benedict and McMahon, 2012; | Mossin, 2018 | Armour et al., 2014; Mell, 2009 |

*Note:* Overall correlation is considered 'direct' where GI interacts with 50 percent or more targets within the respective SDG.

| | |
|---|---|
| Direct Correlation | |
| Indirect Correlation | |
| No Correlation | |

| | | |
|---|---|---|
| Overall Correlation | 50% or more | Direct |
| | Less than 50% | Indirect |
| | | |

**Table 5** A detailed description of the interaction between the SDG 11 targets and GI

| SDG 11: Sustainable cities and communities | | Correlation | Description |
|---|---|---|---|
| 11.1 | By 2030, ensure access for all to adequate, safe and affordable housing and basic services and upgrade slums | Indirect | GI can augment essential services like water and energy. GI elements like green roofs and water harvesting can improve water quality and quantity. Energy use can be reduced by planting trees and building green roofs to regulate the microclimate (Center for Neighborhood Technology, 2010). Introducing the above elements can also lead to upgraded living conditions in informal settlements like slums |
| 11.2 | By 2030, provide access to safe, affordable, accessible and sustainable transport systems for all, improving road safety, notably by expanding public transport, with special attention to the needs of those in vulnerable situations, women, children, persons with disabilities and older persons | Indirect | Employing GI techniques like permeable paving and graded swales to manage stormwater around transport corridors can make land transportation more resilient and safer during flood-like situations. Also, planning ecological tunnels and green bridges in high fauna areas as part of GI can improve road safety and the sustainability of transportation systems |
| 11.3 | By 2030, enhance inclusive and sustainable urbanization and capacity for participatory, integrated and sustainable human settlement planning and management in all countries | Direct | Green infrastructure has been widely argued to be more capable than conventional 'gray' infrastructure in achieving social, environmental and economic objectives (Armour et al., 2014) |

(continued)

**Table 5** (continued)

| SDG 11: Sustainable cities and communities | | Correlation | Description |
|---|---|---|---|
| 11.4 | Strengthen efforts to protect and safeguard the world's cultural and natural heritage | Direct | The cultural heritage element of GI, like forest, land and water, also hold symbolic meanings and religious values. Enhancing, protecting, and conserving GE can aid in safeguarding cultural and natural heritage |
| 11.5 | By 2030, significantly reduce the number of deaths and the number of people affected and substantially decrease the direct economic losses relative to global gross domestic product caused by disasters, including water-related disasters, with a focus on protecting the poor and people in vulnerable situations | Indirect | The introduction of GI can significantly reduce losses due to disasters like floods and droughts (Ely & Pitman, 2014). It aids in water management leading to decreased water loss (green roofs, bio-retention ponds etc.) and can augment water supply by increasing water storage and groundwater recharge. Thus, GI serves to protect vulnerable populations |
| 11.6 | By 2030, reduce the adverse per capita environmental impact of cities, including by paying special attention to air quality and municipal and other waste management | Indirect | GI can play an important role in improving air quality in cities (Ely & Pitman, 2014). For example, vegetation like urban forests and parks can significantly improve air quality |
| 11.7 | By 2030, provide universal access to safe, inclusive and accessible, green and public spaces, in particular for women and children, older persons and persons with disabilities | Direct | Planning and designing green infrastructure involves provisioning green spaces in the form of parks and playgrounds, and improving accessibility |

(continued)

**Table 5** (continued)

| SDG 11: Sustainable cities and communities | | Correlation | Description |
|---|---|---|---|
| 11.a | Support positive economic, social and environmental links between urban, peri-urban and rural areas by strengthening national and regional development planning | Direct | Green infrastructure has the potential to link urban and rural areas by enhancing the environmental and social connections with the introduction of greenways, ecological corridors, and agroforestry, which in turn can strengthen national and regional development |
| 11.b | By 2020, substantially increase the number of cities and human settlements adopting and implementing integrated policies and plans towards inclusion, resource efficiency, mitigation and adaptation to climate change, resilience to disasters, and develop and implement, in line with the Sendai Framework for Disaster Risk Reduction 2015–2030, holistic disaster risk management at all levels | Indirect | The growing awareness of green infrastructure and its contribution to sustainable development and resilience building leads to acceptance and application of green infrastructure as a parallel to mainstream 'gray' infrastructure |
| 11.c | Support least developed countries, including through financial and technical assistance, in building sustainable and resilient buildings utilizing local materials | Indirect | Application of many GI elements like tree planting and community gardens requires minimum technical expertise and money |

tenure; a slum variously lacks space, durability, water, and sanitation; informality implies a lack of formal control over planning, design, and construction" (p. 12). Moreover, unplanned settlements are often situated around areas of high economic value or environmentally sensitive and ecologically sensitive areas, which establishes their connection with green spaces and green infrastructure (see, e.g., Diep et al., 2019; O. Adegun, 2017). The functioning of green spaces as urban can benefit different socio-economic classes of residents within urban areas, "low-income households directly depend on the natural resource environment for their livelihood" (O. B. Adegun, 2019) and consider "green spaces as an amenity" (O. B. Adegun, 2019, p. 79).

Environmental, economic, and social issues inhibit the sustainable development of informal settlements. The issues associated with these settlements have not been systematically addressed and have often been neglected in broader urban and social development practices (Tsenkova et al., 2009). According to UN-Habitat, almost all informal settlements share similar characteristics (United Nations Human Settlements Programme, 2003). Most of them are high density, lack spatial order, and lack public facilities like schools and hospitals. They have poor infrastructure, low build quality, and dilapidated structures (Masele, 2022). The inability of many Global South cities to absorb the current rates of urban growth in terms of providing housing and services leaves informal settlement dwellers vulnerable to environmental, health, and economic risks and results in increasing pressure on urban green infrastructure and ecosystem services.

In the 2030 Agenda directive to *leave no one behind*, all aspects of the built environment requires serious and sustained consideration to achieve this goal.

Green infrastructure has the potential to improve the quality of life and the environment of informal settlements along with delivering the benefits of ecosystem services, resulting in the overall redevelopment or rejuvenation of informal settlements, beyond what is currently offered by upgrading "gray" infrastructure.

While there is evidence of green infrastructure being slowly incorporated into developed urban planning, there is hardly any evidence of green infrastructure being implemented for informal settlements.

Research on green infrastructure is poorly distributed in the Global South, thereby increasing the knowledge gap toward awareness of the importance of the green infrastructure approach. According to Masele (2022), research related to green infrastructure is rarely documented in Africa and Oceania compared to Asia—particularly China—and Latin America. The concept of green infrastructure in the Global North is not limited to academic research, but has found its way into planning policies leading to implementation. The knowledge gap on differences in contexts and drivers in different regions of the Global South constitutes a major challenge for urban sustainability, while the uniqueness of urban issues in the Global South means that knowledge of the Global North is not readily and widely transferable (Nagendra et al., 2018, p. 343).

In Brazil, the concept of green infrastructure has primarily existed in research contexts, but at the government level, while the term is not predominantly reused in legal instruments, various policy documents or institutional frameworks have introduced urban development strategies that coincide with green infrastructure and/or Low-Impact Developments which are based on comparable principles (Diep et al., 2019, p. 565). In India, the green infrastructure concept is still new and is currently being defined by its inclusion in central, state, and sub-regional policies. For example, the 2041 Master Plan for Delhi mentions incorporating blue-green infrastructure (Driver & Mankikar, 2021). In the African context, there is a lack of clear policy, attributed largely to high poverty levels and the prevalence of informality in socio-economic settings (Masele, 2022, p. 9).

Green infrastructure in informal settlements has a significant role in food and nutrition security (Douglas, 2018) "in environmentally sustainable urbanization in developing countries" (Adegun, 2017, p. 22). Floodwater management systems support urban and peri-urban agriculture (Diep et al., 2019). GI contributes to the mitigation of and adaptation to the effects of climate change. Green infrastructure can also contribute to planning and urban design in order to make cities more sustainable (Vásquez et al., 2019, p. 349).

Incorporating green infrastructure in such notoriously developed settlements, however, is a more significant challenge than in more developed urban areas because of insufficient financial and human resources, unclear administrative procedures and policies, a lack of awareness, and insufficient space. As Shackleton et al. (2021) recommend:

> The core ideas and principles of green infrastructure, such as connectivity, multifunctionality, green-grey integration and social inclusion, need to be tailored to the specific situations of the urban areas in Latin America, Africa, and Asia for this purpose with risk management, dealing with poverty and informality being key priorities. (p.132)

Dense informal built environments are especially challenging for implementing green infrastructure. However, GI can be a very flexible tool in densely built environments, e.g., various implementation scales, connectivity, and flexibility of interrelated elements and complementarities with many interlinked urban challenges (M. Ely & Pitman, 2014). GI may require vital project-scale interventions that effectively utilize urban space, for example, roofs (as green roofs), walls (for the vertical garden), and balconies (as a productive garden).

Table 6 lists the various types of informal settlements in Delhi. Parallels of these types of settlements can be drawn from informal settlements with similar characteristics that appear worldwide. Table 7 enlists a sample toolbox that lists the various green infrastructure elements that can be employed in the informal settlements using Delhi as a case example, and their contribution toward various SDGs

## Conclusion

As cities and urban areas continue to grow, interventions are urgently needed to ensure sustainable development. For the sustainable development of urban areas, there needs to be a better balance between gray and green infrastructure. Open areas within the built environment have the potential to be developed into interconnected, multi-functional, multi-scale, natural, and man-made green infrastructure that offer significant social, environmental, and economic benefits. Green infrastructure can be an effective tool to help achieve sustainable development and resilience in the urban environment.

Green infrastructure can play a vital role in benefiting human health and well-being, improving liveability, enhancing prosperity, enhancing climate change resilience, and improving water management and hydrology, urban ecology, and food production.

The potential role of green infrastructure in advancing the 2030 Agenda for sustainable development is not widely considered. As identified through this study, out of the 169 targets of the 17 SDGs, 17 are directly connected to the realization of the 2030 Agenda. An indirect relationship between 44 targets was observed, while 64 percent of targets had no connection with green infrastructure. As per the study, green infrastructure interacts with all 17 SDGs. Green infrastructure has the potential to contribute directly to the realization of SDGs 2, 6, 7, 8, 9, 11, and 15 if implemented appropriately.

This study highlights the need for green infrastructure in informal settlements for the realization of the Agenda 2030 sustainable development objective to *leave no one behind*. There is a need to close the knowledge gap on the importance of green infrastructure, especially in the Global South. More emphasis should be placed on introducing green infrastructure in the design and planning stages rather than being merely an afterthought or retrofit. In the case of informal settlements, green infrastructure should be incorporated on a case-specific basis, keeping in mind the context, unique issues, benefits, and trade-offs.

The primary challenge today is moving green infrastructure into the mainstream, especially where gray infrastructure currently dominates, which requires a broader understanding of the green infrastructure approach and the significant benefits it can deliver (Armour et al., 2014). Furthermore, for green infrastructure

**Table 6** Types of informal settlements in Delhi, their evolution, and their characteristics

| | Type of Informal Settlement in Delhi | Evolution | Characteristic | | | | | |
|---|---|---|---|---|---|---|---|---|
| | | | Open Space (Around Built | Built Structure | Infrastructure | Tenure Security | Environment | Socio-Economic |
| A | Slums (Notified Under Slum Areas Act, 1956) | Encroachment on public and private land with permanent/ semi-permanent hutments | Very Less | Poor | Poor | Partly Secure | Hazardous, polluted | Low |
| B | Squatter Settlements (JJcolony) | Encroachment on public and private land with primarily by temporary hutments | None | Very Poor | None | Not Secure | Hazardous, polluted | Low |
| C | Resettlement Colonies | Relocating squatter & slums to the periphery of the city | Less | Basic | Less than Basic | Partly Secure | Polluted | Low |

(continued)

**Table 6** (continued)

| | Type of Informal Settlement in Delhi | Evolution | Characteristic | | | | | |
|---|---|---|---|---|---|---|---|---|
| | | | Open Space (Around Built | Built Structure | Infrastructure | Tenure Security | Environment | Socio-Economic |
| D | Urban Villages | Rural villages urbanized due to agriculture land acquisition | Less | Basic | Basic or Less than Basic | Secure | Polluted | Middle-low |
| E | Unauthorized Colonies | Developed on agricultural land by violating the Delhi Master Plan | Less | Basic or Poor | Less than Basic | Not Secure | Polluted | Middle-Low |

*Source* Author, with inputs from Ishtiyaq and Kumar (2010)

**Table 7** Toolbox for various GI elements that can be employed in the informal settlements of Delhi as a case example and their contribution toward various SDGs

| | GI Element | SGD | | | | | | | | | | | | | | | | | Application (Informal Settlements) |
|---|---|---|---|---|---|---|---|---|---|---|---|---|---|---|---|---|---|---|---|
| | | 1 | 2 | 3 | 4 | 5 | 6 | 7 | 8 | 9 | 10 | 11 | 12 | 13 | 14 | 15 | 16 | 17 | |
| Green | Public green spaces(Parks, activity areas) | | | ✓ | ✓ | | | | | | | ✓ | | ✓ | | | | | C, D |
| | community garden | ✓ | ✓ | ✓ | ✓ | | | | ✓ | | | ✓ | ✓ | | | | | | A, C, D, E |
| | Tree planting | | | ✓ | | | | | | | | ✓ | | ✓ | | | | | A, C, D, E |
| Grey -Green | Green Roof | | | ✓ | ✓ | | | ✓ | ✓ | | | ✓ | | ✓ | | | | | C, D, E |
| | Green walls | | | ✓ | | | | ✓ | ✓ | | | ✓ | | ✓ | | | | | A, B, C, D, E |
| | Rain gardens | | | ✓ | ✓ | | ✓ | | | | | ✓ | | ✓ | | ✓ | | | A, C, D, E |
| | Bio-swales | | | ✓ | | | ✓ | | | | | ✓ | ✓ | ✓ | | | | | A, C, D, E |
| Grey/ Engineered | Rain barrels | | | | | | ✓ | | | | | ✓ | | ✓ | | | | | A, B, C, D, E |
| | Infiltration trenches | | | | | | ✓ | | | | | ✓ | | ✓ | | | | | A, C, D, E |
| | Pervious pavements | | | | ✓ | | ✓ | | | | | ✓ | | ✓ | | | | | A, C, D, E |

*Source* Author

"to make legitimate contributions to urban sustainability, it must be practiced in a transdisciplinary manner" (Ahern, 2007, p. 282) and be backed with financial investment, long-term policies, and stakeholder engagement.

## References

Adegun, O. (2017). Green infrastructure in relation to informal urban settlements. *Journal of Architecture and Urbanism, 41*. https://doi.org/10.3846/20297955.2017.1296791

Adegun, O. B. (2019). Green infrastructure in informal unplanned settlements: The case of Kya Sands, Johannesburg. *International Journal of Urban Sustainable Development, 11*(1), 68–80. https://doi.org/10.1080/19463138.2019.1565412

Ahern, J. (2007). Green infrastructure for cities: The spatial dimension. In *Cities of the future towards integrated sustainable water and landscape management* (p. 17). IWA Publishing.

Armour, T., Armour, S., Hargrave, J., & Revell, T. (2014). *Cities alive rethinking green infrastructure—Arup*. Arup. https://www.arup.com/perspectives/publications/research/section/cities-alive-rethinking-green-infrastructure

Australian Institute of Landscape Architects. (2012). *Barriers to effective climate change adaptation*. https://www.pc.gov.au/inquiries/completed/climate-change-adaptation/submissions/subdr129.pdf

Barnes, P., O'Hanlon, J., & Matera, M. (2021). *Green > Grey: Challenges, opportunities, and policy implications for delaware green infrastructure landscape* (p. 6). University of Delaware.

Benedict, M. A., & McMahon, E. T. (2001). *Green infrastructure: Smart conservation for the 21st century*. Sprawl Watch Clearinghouse. Retrieved January 26, 2022, from http://www.sprawlwatch.org/greeninfrastructure.pdf

Benedict, M. A., & McMahon, E. T. (2012). *Green infrastructure: Linking landscapes and communities*. Island Press.

Cameron, R. W. F., & Blanuša, T. (2016). Green infrastructure and ecosystem services—is the devil in the detail? *Annals of Botany, 118*(3), 377–391. https://doi.org/10.1093/aob/mcw129

Center for Neighborhood Technology. (2010). *The value of green infrastructure*. Center for Neighborhood Technology. https://cnt.org/sites/default/files/publications/CNT_Value-of-Green-Infrastructure.pdf

Chand Sandhu, S., Naik Singru, R., Bachmann, J., Sankaran, V., & Arnoux, P. (2016). *GrEEEn solutions for livable cities* (0 ed.). Asian Development Bank. https://doi.org/10.22617/BKK157697

Choi, C., Berry, P., & Smith, A. (2021). The climate benefits, co-benefits, and trade-offs of green infrastructure: A systematic literature review. *Journal of Environmental Management, 291*, 112583. https://doi.org/10.1016/j.jenvman.2021.112583

Davies, C., Macfarlane, R., Mcgloin, C., & Roe, M. (2006). *Green infrastructure planning guide*. https://doi.org/10.13140/RG.2.1.1191.3688

Demuzere, M., Orru, K., Heidrich, O., Olazabal, E., Geneletti, D., Orru, H., Bhave, A. G., Mittal, N., Feliu, E., & Faehnle, M. (2014). Mitigating and adapting to climate change: Multi-functional and multi-scale assessment of green urban infrastructure. *Journal of Environmental Management, 146*, 107–115. https://doi.org/10.1016/j.jenvman.2014.07.025

Diep, L., Dodman, D., & Parikh, P. (2019). *Green Infrastructure in Informal Settlements through a Multiple-Level Perspective, 12*(2), 17.

Douglas, I. (2018). The challenge of urban poverty for the use of green infrastructure on floodplains and wetlands to reduce flood impacts in intertropical Africa. *Landscape and Urban Planning, 180*, 262–272. https://doi.org/10.1016/j.landurbplan.2016.09.025

Dovey, K., & King, R. (2011). Forms of informality: Morphology and visibility of informal settlements. *Built Environment, 37*. https://doi.org/10.2148/benv.37.1.11

Driver, B., & Mankikar, S. (2021). *Blue-green infrastructure: An opportunity for indian cities*. Observer Reserach Foundation.

du Toit, M. J., Cilliers, S. S., Dallimer, M., Goddard, M., Guenat, S., & Cornelius, S. F. (2018). Urban green infrastructure and ecosystem services in sub-Saharan Africa. *Landscape and Urban Planning, 180*, 249–261. https://doi.org/10.1016/j.landurbplan.2018.06.001

Ely, M., & Pitman, S. (2014). *Green infrastructure project* (p. 380). Botanic Gardens of South Australia Department of Environment, Water and Natural Resources.

European Commission. (2013). *Green Infrastructure (GI)—Enhancing Europe's Natural Capital*. https://ec.europa.eu/environment/nature/ecosystems/docs/green_infrastructures/1_EN_ACT_part1_v5.pdf

European Commission. Directorate General for the Environment. (2014). *Building a green infrastructure for Europe*. Publications Office. https://data.europa.eu/doi/10.2779/54125

Filho, W. L., Tortato, U., & Frankenberger, F. (2019). *Universities and sustainable communities: Meeting the goals of the agenda 2030*. Springer Nature.

Hansen, R., & Pauleit, S. (2014). From Multifunctionality to multiple ecosystem services? A conceptual framework for multifunctionality in green infrastructure planning for urban areas. *Ambio, 43*(4), 516–529. https://doi.org/10.1007/s13280-014-0510-2

Hansen, R., Rall, E., Chapman, E., Rolf, W., & Pauleit, S. (Eds.). (2017). *Urban green infrastructure planning: A guide for practitioners*. GREEN SURGE. http://greensurge.eu/working-packages/wp5/

Ishtiyaq, M., & Kumar, S. (2010). *Typology of informal settlements and distribution of slums in the NCT of Delhi* 14.

Kramer, M. G. (2014). *Enhancing sustainable communities with green infrastructure* (p. 66). EPA Office of Sustainable Communities. https://www.epa.gov/sites/default/files/201608/documents/green-infrastructure.pdf

Loftness, V., & Haase, D. (Eds.). (2013). *Sustainable built environments*. Springer. https://doi.org/10.1007/978-1-4614-5828-9

Maes, M., Jones, K., Toledano, M., & Milligan, B. (2019). Mapping synergies and trade-offs between urban ecosystems and the sustainable development goals. *Environmental Science & Policy, 93*, 181–188. https://doi.org/10.1016/j.envsci.2018.12.010

Masele, W. (2022). Exploring urban green packages as part of Nature-based Solutions for climate change adaptation measures in rapidly growing cities of the Global South. *Journal of Environmental Management, 310*. https://doi.org/10.1016/j.jenvman.2022.114786

Mell, I. (2009). Can green infrastructure promote urban sustainability? *Proceedings of the Institution of Civil Engineers-Engineering Sustainability, 162*, 23–34. https://doi.org/10.1680/ensu.2009.162.1.23

Mossin, N. (2018). *An architectural guide to the UN 17 Sustainable Development Goals.* Institute of Architecture and Technology, KADK, The Danish Association of Architects, The UIA Commission on the UN Sustainable Development Goals. https://uia2023cph.org/wp-content/uploads/2022/05/AN_ARCHITECTURE_GUIDE.pdf

Nagendra, H., Bai, X., Brondizio, E. S., & Lwasa, S. (2018). The urban south and the predicament of global sustainability. *Nature Sustainability, 1*(7), 341–349. https://doi.org/10.1038/s41893-018-0101-5

Natural England. (2009). *Green infrastructure guidance.*

Schröter, D. E., Wörlen, M., Röber, D. J., Dreiseitl, H., Kishnani, N., Yok, T. P., Cossu, G., Ng, C., Jr, J. L. W., Marks, A., Noiva, K., Rawoot, S., Rosenthal, J. K., & McGlynn, E. (2016). *Strengthening blue-green infrastructure in our cities* (p. 174). Ramboll Foundation.

Shackleton, C. M., Cilliers, S. S., Davoren, E., & du Toit, M. J. (Eds.). (2021). *Urban ecology in the global south.* Springer International Publishing. https://doi.org/10.1007/978-3-030-67650-6

*The Sustainable Development Goals Report 2022.* (2022). United Nations. https://unstats.un.org/sdgs/report/2022/The-Sustainable-Development-Goals-Report-2022.pdf

Titz, A., & Chiotha, S. (2019). Pathways for sustainable and inclusive cities in Southern and Eastern Africa through urban green infrastructure? *Sustainability, 11*, 2729. https://doi.org/10.3390/su11102729

Town and Country Planning Association. (2004). *Biodiversity by Design—Town and Country Planning Association.*

Tsenkova, S., Potsiou, C., & Badyina, A. (2009). *Self-made cities: In search of sustainable solutions for informal settlements in the United Nations Economic Commission for Europe region* (United Nations, Ed.). UNECE Information Service.

United Nations. (2015). *Transforming our world: The 2030 Agenda for Sustainable Development.* https://sdgs.un.org/2030agenda

United Nations Human Settlements Programme. (Ed.). (2003). *The challenge of slums: Global report on human settlements.* Earthscan Publications.

Vásquez, A., Giannotti, E., Galdámez, E., Velasquez, P., & Devoto, C. (2019). Green infrastructure planning to tackle climate change in Latin American Cities. In *urban climates in Latin America* (pp. 329–354). https://doi.org/10.1007/978-3-319-97013-4_13

# Solar Shading Design and Implementation in UK Housing as a Tool for Advancing Sustainable Development

Claire Brown

## Introduction

Core points:

- Housing is an important topic when addressing the issue of climate change—global greenhouse gas emissions must be reduced by 45% by 2030.
- Housing is a key determinant of health. Overheating in homes in summer will become increasingly problematic as the planet warms.
- Housing designers need to think about meeting multiple goals, and the SDGs offer a framework for them to do this.
- This chapter looks at how the technology of solar sharing can reduce overheating and cut climate emissions, helping meet three goals at once: SDG 3, 11, and 13.

Urgent action is needed to recalibrate how climate change is dealt with. It is no longer a side issue that some scientists are discussing. Instead, it is a global issue that has far-reaching consequences. The impacts of climate change are being felt across the planet, from rising sea levels in Hawaii to increased temperatures in India. These effects directly impact the ability of people to lead their lives.

C. Brown (✉)
Department of Mechanical, Aviation and Civil Engineering, Faculty of Science and Engineering, University of Manchester, Manchester, UK
e-mail: claire.brown-3@manchester.ac.uk



T. Walker et al. (eds.), *The Role of Design, Construction, and Real Estate in Advancing the Sustainable Development Goals*, Sustainable Development Goals Series, https://doi.org/10.1007/978-3-031-28739-8_5

## Scope

To feel safe and secure in our homes and dwellings, we need to achieve a suitable level of thermal comfort. In a changing climate, this can be a challenge. The premise of this chapter is to consider how solar shading can be part of the solution. This calls upon academic research and reports from industry and professional bodies. The United Kingdom (UK), as a case study example, offers interesting insights into how technology can be employed in a geographic setting not ordinarily considered at risk from solar gain.

## Hypothesis

More global deaths are expected because of the impacts of a changing climate, more specifically, because of predicted overheating, with some reports suggesting an increased mortality of 257% by the 2050s (Lee & Steemers, 2016; Lee et al., 2015). Increasing mortality rates are a serious issue. The UK is predicted to reach approximately 6000 deaths per annum by 2050 (Grussa et al., 2019) as a result of changing and increasing temperatures. The current rate of application of solar shading to reduce climate impacts is limited in the mainstream domestic housing sector, even though a body of work supports its implementation. To meet climate targets and reduce mortality, overheating risks must be addressed. To reach current climate targets, there needs to be a 45% reduction in carbon dioxide emissions from 2010 levels by 2030 (IPCC, 2021) and housing makes up one-fifth of carbon emissions, the next highest sector after transportation. Solar shading can be part of the solution. Properties in a temperate climate, such as the UK, are well placed to employ shading technology to reduce the associated human health risks of overheating. However, at present, this technology is not widely adopted in the UK. This chapter explores why this might be the case and offers some design options for implementing this change. This chapter will explore how solar shading can play a role in climate change adaptation and mitigation plans for the UK.

This chapter focuses on resilient design solutions around the implementation of solar shading for new build housing. The focus is on design principles: strengths and weaknesses of various measures that can be used in homes to reduce the risk from the impacts of climate change. When considering how to combat the effects of climate change in the Built Environment, specifically housing, we can begin by looking at options to reduce the impact of solar gain. Knowledge of these options is essential for the engineer, owner, or occupant.

## Methods

This chapter uses a variety of data sources and literature databases to compare and review the research about solar shading as a method to combat climate change as part of sustainable development. This information will be useful to a range of readers on how best to apply the technology to achieve the UN Sustainable Development Goals (SDGs). These are outlined in Table 1.

## Literature Review

This section focuses on material that has been previously published. This includes literature in the public, private and academic sectors on the issue of sustainable development and the potential for solar shading to help support climate change targets. The databases used include Elsevier, Springer, and Google Scholar. This data method gathering is discussed in the next section.

## Sustainable Development and the SDGs

The idea of sustainable development and one of the key definitions within wider governance communities was published in the Brundtland Report in 1987. This highlighted the need for a consideration of both the ecological and biological importance (Brundtland, 1987) and the consideration that this was about global development. The Brundtland Commission stated its definition as:

> To ensure the growth of human progress through development without bankrupting the resources of future generations. (Brundtland, 1987)

This definition brings to center stage the idea of the development of other than economic goals. Following this, the 1992 UN Conference on Environment and Development (UNCED) took place in Rio de Janeiro, Brazil, and brought forward the 1987 definition of sustainable development (Brundtland, 1987). This session sparked a discussion of future impacts and, in part, led to Agenda 2030, which formed the basis for the United Nations Sustainable Development Goals (UN SDGs). The global push to effect change is supported by other international agreements, such as the Kyoto Protocol, the Montreal Protocol, the Sendai Framework, and the Paris Agreement. The details of these agreements are outlined in Table 2.

The UN SDGs are a series of global targets intended to support the 2030 Agenda for addressing poverty and injustice. These seventeen interlinked targets set out the conditions required to meet the 2030 Agenda. Similar tools for creating change are often used in business management. Managing change at the global

**Table 1** Key papers from the literature that consider how solar shading can be part of the solution

| Author and Year | Title | Key Findings | Country of Case Study |
|---|---|---|---|
| A.-M. M. Makantasi and A. Mavrogianni (2016) | Adaptation of London social housing to climate change through retrofit: a holistic evaluation approach | UK Social housing is at risk of overheating due to a changing climate. Reducing overheating hours by 28% (Makantasi & Mavrogianni, 2016) | UK |
| van Hooff, T et al. (2014) | On the predicted effectiveness of climate adaptation measures for residential buildings | Solar shading is an indicator of the number of overheating hours experienced by a case study unit | Netherlands |
| Tettey, U et al. (2019) | Design strategies and measures to minimize operation energy use for passive houses under different climate scenarios | Future Climate change leads to a reduction in heating demand and overheating risk goes up, but solar shading can be used to mitigate this | Sweden |
| Tillson, A et al. (2013) | Assessing impacts of summertime overheating: Some adaptation strategies | UK building stock is not designed currently to withstand climate change impacts including overheating | UK |
| Mavragionni, A et al. (2015) | Urban social housing resilience to excess summer heat | A London case study found that shading can significantly help in reducing overheating issues | UK |

(continued)

**Table 1** (continued)

| Author and Year | Title | Key Findings | Country of Case Study |
|---|---|---|---|
| Mavrogianni et al. (2017) | Inhabitant actions and summer overheating risk in London dwellings | An assessment of London dwellings found that current homes are currently at risk of overheating | UK |
| Wright, A., and Venskunas, E. (2022) | Effects of Future Climate Change and Adaptation Measures on Summer Comfort of Modern Homes across the Regions of the UK | With the application of solar shading, the risk is reduced significantly for UK case study homes | UK |

**Table 2** Key UN-based agreements and policies in the context of the built environment

| Name | Year | Detail |
|---|---|---|
| Montreal Protocol | 1987 | Multilateral agreement from the UN on the reduction in the use of substances that deplete the Ozone Layer. https://www.unep.org/ozonaction/who-we-are/about-montreal-protocol |
| Kyoto Protocol | 2005 | Ratifies the UN framework convention on Climate Change to limit carbon emissions and lower levels within agreed targets. https://unfccc.int/kyoto_protocol |
| Sendai Framework for Disaster Risk Reduction | 2015 | Focusing on natural disaster prevention in all nations and how to eliminate or mitigate the risks faced. A 15-year framework plan. https://www.undrr.org/publication/sendai-framework-disaster-risk-reduction-2015-2030 |
| Paris Agreement | 2015 | An international agreement on climate change. The aim is to reduce global warming. https://unfccc.int/process-and-meetings/the-paris-agreement/the-paris-agreement |

scale is an onerous task and when the change involves issues of global poverty and inequality, it can also be especially challenging. However, the UN SDGs and the 2030 Agenda press forward despite resistance from some areas.

Global agreements can be seen as unifying agents of change but depend greatly on the ability of signatories to follow agreements. This difficulty in achieving meaningful outcomes can be addressed using milestones and individual targets. For the UN SDGs, this is a valuable way of being able to track change for its seventeen goals.

The risk to human health can be aligned to the framework of the United Nations Sustainable Development Goals: SDG 3 (Good health and Well-being), SDG 11 (Sustainable Cities and Communities), and SDG 13 (Climate Action). Ensuring the impacts of climate change are adequately addressed includes the need to ensure the creation of sustainable and resilient homes. Reducing deaths from overheating is a clear focus of UN SDG 13. Overheating is a human health risk that requires urgent attention. By reducing the impact of solar glare on homes the issue can begin to be addressed.

Taking the lead from three of the 17 United Nations Sustainable Development Goals (UN SDGs), this chapter will consider how the built environment within the context of a changing climate can be supported by design changes within construction, particularly for housing and by the use of solar shading. Three SDGs and associated selected targets are selected from a potential list of 169 targets (Fuldauer et al., 2022). Three are the focus of this chapter. Table 3 provides details of these SDGs and related targets.

The targets illustrated in the table are all key in ensuring that Agenda 2030 is met. Here, we look at how solar shading is part of the solution and a mechanism to achieve these targets. This is within a defined framework of addressing issues in sustainable development: how can the built environment sector move toward meeting the targets set within Agenda 2030 and progress both adaption and mitigation measures for climate resilience?

In the early 1970s, the first oil crisis significantly impacted the UK, particularly as a focus and link between energy consumption and everyday life (Bouzarovski, 2020). The crisis represented a critical moment that highlighted the dependence of the country on fossil fuels. The sudden reduction across the globe served to illustrate that fossil fuels were not an infinite resource of cheap, reliable energy. A power outage in Europe in 2006 resulted in over 15 million people experiencing a blackout (Hitchin, 2014). Certainty of supply and the ability to get energy services into a home is essential services relied on by individuals, families, and especially vulnerable people who are reliant on energy to manage medical conditions.

**Table 3** UN SDGs 3, 11, and 13 applicable to solar shading and the built environment

| ICON | Headline | Target | Link to Solar Shading |
|---|---|---|---|
|  | 3 – Ensure healthy lives and promote well-being for all at all ages | 3.d. Strengthen the capacity of all countries, in particular developing countries, for early warning, risk reduction, and management of national and global health risks | Climate change is a global health risk, especially in the context of overheating. Solar shading can be employed within the built environment to help combat this |

(continued)

**Table 3** (continued)

| ICON | Headline | Target | Link to Solar Shading |
|---|---|---|---|
| | 11 – Make cities and human settlements inclusive, safe, resilient, and sustainable | 11.b. By 2020, increase the number of cities and human settlements adopting and implementing integrated policies and plans towards inclusion, resource efficiency, mitigation and adaptation to climate change, resilience to disasters, and development and implementation, in line with the Sendai Framework for Disaster Risk Reduction 2015–2030, holistic disaster risk management at all levels (UN, 2018) | Adaption to climate change includes technology and design relating to homes. New homes can be designed to include suitable levels of shading to prevent overheating risks |
| | 13 – Take urgent action to combat climate change and its impacts (Nations, 2021) | 13.2 Integrate climate change measures into national policies, strategies, and planning | The built environment needs to address a changing climate through updated design philosophy and construction of housing units |

Energy demand and energy vulnerability are an issue of concern both in the UK and globally. Specific socio-economic groups and other vulnerable populations are particularly susceptible to loss of electricity (Pye et al., 2021). Policy changes related to supply and demand and how to combat undersupply issues have been implemented. Conservation of energy (Part L) of the building regulations came into play in 1985 as part of building compliance, specifically for power conservation, and provides the framework for energy efficiency in buildings (MHCLG, 2021).

The changing climate needs to be addressed in both the UK and globally. The effects of climate change are witnessed in the UK, including increased risks of flooding, storms, and heat waves. Passive cooling design can be part of the plan to alleviate the associated risks, especially overheating. This strategy focuses on removing excessive heat from the dwelling or home (Santamouris, 2016). The removal of heat or prevention of overheating is essential in helping to reduce the energy load added to the home through mechanical cooling or purge systems, thereby reducing cost and the load on the electricity grid. While heat remediation is currently a focus in Sub-Saharan and South Asian urban centers (Waite et al., 2017) on mass, this is also an issue for the UK, especially for areas where a significant summer temperature increase is expected. The UK Met Office provides United Kingdom Climate Projections findings which highlight both 2 °C and 4 °C (Met Office, 2021) mapped increases for the UK. This includes London, certain areas in the southeast of the UK, and urban centers in the West Midlands (Kennedy-Asser et al., 2022).

This is particularly important when we look at the level of increased temperatures that lead to summertime overheating (Adekunle & Nikolopoulou, 2016) for these areas.

The principal issues for the UK will be around the changing climate and thermal comfort. Research suggests that heating load and demand will be reduced, but not entirely replaced, by a need to cool (Collins et al., 2010). Mitigation and adaption will need to focus on not one, but two temperature gauges, adjusting the thermal environment to acceptable levels for human comfort and well-being. Fanger (1973) concluded that creating optimal thermal comfort for a group is a challenge. While our knowledge has increased since the factors influencing thermal comfort are even more complex under changing climate conditions. The UK Chartered Institute of Building Service Engineers (CIBSE) also reports that there are several factors to consider when thinking about thermal comfort (Lattimore, 2020).

These are:

1. Metabolic rate.
2. Clothing.
3. Air temperature.
4. Mean radiant temperature.
5. Airspeed.
6. The humidity of the space.

These are all considerations for what level of thermal comfort should be designed for. In pushing the sustainable development ethos, the built environment needs to address a changing climate that will impact people differently. Personal thermal comfort might not be able to be achieved for all individuals, but it can be addressed to some degree by design changes and retrofit options.

The legislation means that there are legally binding targets that need to be achieved within the UK. Efficiency and climate resilience are part of the mix. UK housing is known to be inefficient in terms of energy consumption. A changing climate will significantly impact how people can adapt. Human health risk factors are key drivers for driving the change. The UN SDGs can guide meaningful changes at the global level, however, until local and country-level policies are implemented, change comes down to the householder or the property owner. The idea of solar shading is to reduce the amount of solar gain from the sun that reaches the building. The solar gain factor 'g' is used to express the proportion of heat gain within the structure resulting from total solar irradiance incident on the outside surface of the glazing. Practically, this has the same value as the solar heat gain coefficient (SHGC), which is a more popular measure in some areas of the world.

The point at which solar gain directly enters the home is concentrated around the window glazing. The 'g' value depends on the specific glazing used. This can be seen below:

$$\text{g-value} = \text{total solar heatgain}/\text{incident solar radiation}$$

The point at which solar gain directly enters the home is concentrated around the glazing and windows specified. The effect of solar heat gain is influenced by the cleanliness, surface reflectance, and type of glazing found in the building fabric. The technical feasibility should include all possibilities at the design stage (Cheshire, 2020) and include any climate resilience and support for sustainable development. The most effective way to manage overheating is to prevent sunlight from reaching the windows. The application of shade to prevent excess heat is a realistic solution for many domestic properties. This is especially true for properties that need to address issues of climate change mitigation and adaptation. Parameters such as depth, size of the building, and building orientation of homes are important when considering the potential risk for overheating. In the case of solar shading, the type and number of glazing materials are important considerations, but for mitigation, it is the design that is of paramount importance.

Understanding how heat can enter a dwelling is essential. Heat can be gained through both natural and artificial measures, for example, the heat gain from the sun versus the internal gains from energy-consuming services. All of this contributes to the amount of heat gain of a property. When overheating becomes an issue, it must be addressed. Solar strategies for homes can provide the answer to overheating (Rahif et al., 2022) within the home. The following section considers solar strategies in more detail.

## Solar Shading—A Review

It has long been recognized that solar strategies, particularly domestic properties, have not been fully explored (Baborska-Narożny et al., 2017). This is especially true in the context of a changing climate. Solar strategies and green buildings have been incompletely addressed (Fassbender et al., 2022). Building integration is not widely researched at present (Vassiliades et al., 2019). It is an issue that needs to be considered and it is especially important in the context of meeting the Agenda 2030 deadline. Many different climate mitigation technologies have been mentioned in the research (Fassbender et al., 2022). However, technology is not implemented fully at scale in domestic buildings. This may explain why the uptake has not been more widespread. We do see that the technology is used in specific climate zones and specific applications. The inclusion of solar shading technology has continued to the present day (Frattolillo et al., 2020). Much more needs to be done, especially to meet UN SDGs. The number of people likely to be affected by overheating risks continues to increase each year.

Solar shading is not complicated to install, but for the best results, it needs to be included at the design stage for a new build. It can also be retrofitted. For some homes, this might consist of a curtain or a blind. For new homes, consideration of solar gain at the design stage is of utmost importance. Retrofitting presents an opportunity to make fabric and other technological changes to the structure. Climate change presents challenges that need to be addressed for older properties. The UK Met Office reports, "One billion face heat-stress risk from 2 °C rises" (Madge, 2021). Therefore, it is important to discuss climate risks and why this issue is covered by multiple UN SDGs.

Climate risk is covered within the UN SDGs. Heat risk and the potential for overheating are the subjects of research in the UK. The subject of blinds to combat the issue of solar gain and solar glare appears in several publications. Installing blinds in the home is a relatively easy process, which the owner or occupant can undertake on their own. Integrated systems or standalone units can also be specified at the design stage. Curtains and blinds can also be of benefit in reducing heat loss during cooler months.

Solar shading can be divided into two categories—permanently fixed shading and moveable shading (which can be operated by the tenant or occupant). In most cases, the solar shading strategy will be assigned and developed using a dynamic thermal model to assess the likelihood of any potential overheating risk because of solar gain. Simple and inexpensive interventions can mitigate the impact of any risks due to climate change, such as blinds and solar screens. Table 4 lists some of the most common design features for reducing solar gain.

**Table 4** A range of solar shading options, including ease of installation and maintenance

| System | Ease of Application | Ease of Maintenance | Thermal reduction | Allow View |
|---|---|---|---|---|
| Reflective Blinds | Window | Easy | Yes | Limited |
| Venetian Blinds | Window | Easy | Yes | Limited |
| Moveable Blinds | Window | Easy | Yes | Limited |
| Light Shelves | Window | Easy | Yes | Yes |
| Translucent Louvres | Window | Difficult | Yes | Limited |
| Solar Screens | Window | Difficult | Yes | Limited |
| Holographic Films | Inside double glazing | Easy | Potential | Yes |
| The active modular glazing panel | Window | Easy | Yes | Yes |

*Source* Adapted from (Wong, 2017)

The solar shading performance of a building can be assessed using various methods, which are readily available in manual design tools or dynamic simulation software.

- EnergyPlus was used by Mavrogianni et al. in their research in 2012 on building characteristics and indoor temperatures. Housing stock and archetype units were assessed to understand performance in a changing climate (Mavrogianni et al., 2012).
- CIBSE TM59 specifically addresses the issue of overheating within domestic properties (Grussa et al., 2019).
- Guidance from CIBSE in the form of Environmental Design Guide A (Gupta & Gregg, 2020) has been used to provide metrics for temperature levels in assessments.

With any system used as an assessment tool, there will be strengths and weaknesses. Modeling, guidance documents, and virtual environments can all be valuable tools in promoting climate resilience during the design stage. However, the strength of a tool lies in reliability in highlighting issues around risk. As a tool for advancing sustainable development, this can be an asset to the client, contractor, and end-user.

Computer design and computer-based simulation tools can be useful in assessing sustainable development options and needs in the design context (Wong, 2017). One area that is used increasingly in dynamic building simulation, is the assessment of overheating risk. Understanding how solar shading is part of the solution to an overheating risk is important. Modeling, and particularly, dynamic thermal modeling can be especially useful in supporting any proposals for change. Building Performance Evaluation is an essential process in understanding the implementation of any proposed technology (Sharpe et al., 2018). This includes solar elements. New buildings can incorporate elements for managing solar gain at the design stage. A correct understanding of the structure needs to be recognized. Purpose and occupants are important in this. Knowing who the tenants of the structure will be and the implications of overheating must not be underestimated. Vulnerable people are particularly susceptible to fluctuating temperatures. It is also important to recognize that approximately 50% of the world's population lives in cities and urban settings (Bastin et al., 2019).

Transitioning to a climate-resilient housing system and one where sustainable development is at the core is essential. Resilience comes from the ability to respond to stress. A climate-resilient design response is needed to ensure that what is built is appropriate for the changing climate (Widera, 2021). The

UK Department for Business, Energy, and Industrial Strategy (BEIS) agrees that buildings need to be fit for the future. BEIS promotes high energy efficiency standards, appropriate for future buildings, within the context of sustainable development (BEIS, 2021). Climate resilience means having a safe place to call home that is not going to overheat, flood, or be damaged by storms. This includes housing, climate-resilient homes will have to be part of a sustainable future.

New social housing in the UK and the scale of development are affected by changing standards and requirements, and disruption to regulations creates uncertainty in the market. This can, in some cases, result in inertia. Changes in technology and innovation, particularly involving climate change, can result in slow adoption. Feedback is a vital element when solar shading technology has been included in any development (Gupta et al., 2018).

The following section will consider the key findings from the research. Urban overheating is a real risk to energy balance within cities across the globe (Su et al., 2021). The balance between heating and cooling is an interesting one, with research pointing to how a tipping point will be reached. This is when the cooling demand exceeds the heating demand. In their 2016 paper, Makantasi and Mavrogianni reported that while solar shading will reduce cooling demand by 28%, there is likely to be an increase in heating demand of 19.5% (Makantasi & Mavrogianni, 2016) as a result of reduced solar gain in the heating season. Other research has concluded that moveable solar shading can be of great benefit.

Van Hooff and their team in the Netherlands found that when the solar gain of a window is 150 W/m$^2$ or larger the solar shading has a large impact on overheating in a dwelling (van Hooff et al., 2014). Again, support for adjustable and adaptive facades can be considered a useful adaptation element for buildings (Valitabar et al., 2022). The ability to change and update the structure depending on the season can potentially help people with gaining control over their environment and improve thermal comfort.

Sustainable development will not see a single technology or design feature being implemented. A range of techniques will be used to affect the grid consumption of the dwelling. Tettey et al. (2019), found that energy demand, including the solar shading of windows, could be reduced by 40–51% in a case study of archetype dwellings. Taking climate action will require a multilevel and multifaceted approach to succeed. This need for appropriate action will differ in terms of the size and scale of development and the construction techniques used. What is critical to note is the ability to reduce the overheating risk.

When overheating risk is considered, the link can be made directly to the UN SDGs, 3, 11, and 13. The targets for this related to overheating risk are listed below:

- 3.d. Strengthen the capacity of all countries developing countries, for early warning, risk reduction, and management of national and global health risks,
- 11.b. By 2020, increase the number of cities and human settlements adopting and implementing integrated policies and plans toward inclusion, resource efficiency, mitigation and adaptation to climate change, resilience to disasters, and development and implementation, in line with the Sendai Framework for Disaster Risk Reduction 2015–2030 (UN, 2018), holistic disaster risk management at all levels and
- 13.2 Integrate climate change measures into national policies, strategies, and planning.

Overheating is a global health risk, which can be addressed through Solar Shading as a standalone technology or as part of a wider group of technologies. Solar shading is capable of supporting a reduction in overheating risk levels to 6% (Tillson et al., 2013). This can help to reduce the risk and support UN SDG target 3.d. Adaptation will be necessary to prevent human health risk levels from being elevated because of climate change. Overheating and the risk of human harm can be reduced through solar shading (Wright & Venskunas, 2022).

A joint strategy for addressing 11.b UN SDG to address the risks associated with climate change needs to be considered. Mavrogianni et al. found that shading and night cooling was better than all-day rapid ventilation (Makantasi & Mavrogianni, 2016; Mavrogianni et al., 2015), but issues of air pollution could lead to other human health risks higher than overheating alone. Homes are designed within a remit set by the client and laid down by local requirements for planning and building regulations. Critical links between development and the SDG goals and targets are linked to the ability for change and policy support (Fuldauer et al., 2022).

Target 13.3 for the UN SDGs recognizes a need for change at the governance level. The ability to change and adapt is important, particularly in the context of climate change and overheating. Recognizing compliance is also important. Solar shading is not a recent technology. Many cities worldwide already experience extreme temperatures; from these examples, the UK can learn but needs a mechanism to do this. Green tech can play a role in reducing the impacts of climate change (Bortone et al., 2022), but policy needs to support the role that it can play when considering the likely impacts of the changing climate on people.

Building technology and specific examples can lead to unintended consequences and an increase in consumption rather than a reduction. An example is when people believe they live in an energy-efficient home and can therefore be careless with energy usage because it is deemed a well-performing dwelling. Unintended consequences of various technologies should also not be ignored; anything mechanized will need maintenance or energy to power it. Education about how to manage technology would be a welcome addition here. The improvement of Homes User Guides (HUGs) for new dwellings improves occupant understanding of how to use technology (Baborska-Narożny et al., 2017).

Detailed discussions around the benefits of solar shading exist in the commercial sector, especially regarding office buildings. Research has shown that when solar shading was implemented in offices, savings could be made concerning energy consumption and reduced carbon emissions. Buildings where a cooling regime was implemented and where an external automatic shading control system was in place found savings of 66% for London (Littlefair et al., 2010). In the same research, differences were noted for the office units surveyed for fixed shading systems. For a fixed system, in London, a saving was noted of a 5% energy cost reduction. Research from the Netherlands, which focused on the domestic market found that savings concerning energy can be made from installing solar shading. Importantly, it recognized that combined shading regimes found a saving of 30% with associated adaptive thermal comfort systems in their case study (Alders, 2017). Australian research has also found that mitigation and adaptation measures which include solar shading can supply a significant reduction in associated energy costs. Reducing future cooling demand by the implementation of shading in the Sydney case study found a reduction of 31.3 kWh/m2 (70%) (Garshasbi et al., 2020). The search for equivalent data for the UK is limited.

Mitigation of climate change impacts is important for the construction of homes in the UK. Increased temperatures will see a negative impact on occupants. The UN SDGs include specific details regarding this issue.

- 3.d. Strengthen the capacity of all countries, in particular developing countries, for early warning, risk reduction, and management of national and global health risks,
- 11.b. By 2020, increase the number of cities and human settlements adopting and implementing integrated policies and plans toward inclusion, resource efficiency, mitigation and adaptation to climate change, resilience to disasters, and development and implementation, in line with the Sendai Framework

for Disaster Risk Reduction 2015–2030 (UN, 2018), holistic disaster risk management at all levels and
- 13.2 Integrate climate change measures into national policies, strategies, and planning.

All countries no matter what their climate will see benefits in addressing global health risks, such as overheating. Plans and policies within local government can strengthen the response that is needed to protect human life. It is important not to underestimate what the provision of solar shading technology can do to mitigate the effects of overheating. Reducing the risk to human health will have benefits for everyone, especially those vulnerable communities who are likely to experience the very worst effects of climate change.

The power to change and adapt needs to be incorporated more into construction methods. If an overhang can be included in the design of a property, it means that the heat gain in summer is reduced, and the need for higher-performing ventilation might be offset. A careful balance within the design is what is needed. Reducing solar gain for overheating risk is a promising technology that has the potential to address the issues of a changing climate and reduce solar gain for UK housing. Sustainable construction practices can include solar shading, which would support a reduction in heat gain. However, we are not seeing enough examples. Another heatwave in the UK may be the trigger that kick-starts the change we need. The following are key takeaways from this chapter.

- Solar shading is important as a construction tool for sustainable development.
- A variety of technologies can be installed during construction or potentially as a retrofit solution.
- A changing climate may require replacement technology if significant temperature changes occur.
- House design and risk of exposure to excess heat must be part of any assessment to understand which solar shading technology should be deployed.

## Discussion

Sustainable development and aligning it with the UN SDGs is achievable from the evidence from many studies. The evidence points toward solar shading being a wholly suitable method for reducing heat gain and therefore reducing the need for energy for cooling for some applications. Research has shown that we understand how this can be implemented in several climates. Building technology exists at various scales. What is missing is the drive to increase the performance of buildings. Regulation and compliance are crucial drivers for the construction industry. To ensure change happens, the regulatory framework in the UK needed to change with the level of commitment and speed to fully address the impacts of climate change.

The UN SDGs are not a legal framework, they are a set of clear guidelines for sustainability, so the question remains as to how to push forward with meeting the UN SDG goals and promote the use of solar shading to alleviate overheating and excessive heat gain. Solar shading is one way of addressing the issues of climate change, particularly around overheating, but other measures are needed for a significant reduction in carbon emissions to be achieved. Ventilation and energy-saving devices are also important, and solutions should not be looked at in isolation. A collective approach across the sector of construction needs to be unified in the push for improvements. Making the necessary sectoral changes to support sustainable development. Reducing solar gain in homes is needed now, but in 2030 it will be even more necessary.

## References

Adekunle, T. O., & Nikolopoulou, M. (2016). Thermal comfort, summertime temperatures, and overheating in prefabricated timber housing. *Building and Environment, 103*, 21–35. https://doi.org/10.1016/j.buildenv.2016.04.001

Alders, E. E. (2017). Adaptive heating, ventilation, and solar shading for dwellings. *Architectural Science Review, 60*(3), 150–166. https://doi.org/10.1080/00038628.2017.1300132

Baborska-Narożny, M., Stevenson, F., & Grudzińska, M. (2017). Overheating in retrofitted flats: Occupant practices, learning and interventions. *Building Research and Information, 45*(1–2), 40–59. https://doi.org/10.1080/09613218.2016.1226671

Bastin, J.-F., Clark, E., Elliott, T., Hart, S., Van Den Hoogen, J., Hordijk, I., Ma, H., Majumder, S., Manoli, G., Maschler, J., Mo, L., Routh, D., Yu, K., Zohner, C. M., & Crowther, T. W. (2019). Understanding climate change from a global analysis of city analogues. *PLoS ONE, 14*(7), e0217592. https://doi.org/10.1371/journal.pone.0217592

BEIS. (2021). *Heating and buildings strategy*. OGL https://www.gov.uk/government/publications/heat-and-buildings-strategy

Bortone, I., Sakar, H., & Soares, A. (2022). Gaps in Regulation and policies on the application of green technologies at household level in the United Kingdom. *Sustainability, 14*(7), 4030. https://doi.org/10.3390/su14074030

Bouzarovski, S. (2020). Transforming urban energy demand: A timely challenge. *Frontiers in Sustainable Cities, 2*. https://doi.org/10.3389/frsc.2020.00029

Brundtland, G. H. (1987). *Report of the world commission on environment and development: Our common future*. United Nations General Assembly, Issue. https://sustainabledevelopment.un.org/content/documents/5987our-common-future.pdf

Cheshire, D., & Godefroy, J. (2020). *CIBSE Guide L*. CIBSE. https://www.cibse.org/knowledge-research/knowledge-portal/guide-l-sustainability-2020

Collins, L., Natarajan, S., & Levermore, G. (2010). Climate change and future energy consumption in UK housing stock. *Building Services Engineering Research and Technology, 31*(1), 75–90. https://doi.org/10.1177/0143624409354972

Fanger, P. O. (1973). Assessment of man's thermal comfort in practice. *Occupational and Environmental Medicine, 30*(4), 313–324. https://doi.org/10.1136/oem.30.4.313

Fassbender, E., Ferdinand, L., Andreas, H., Thomas, A., & Claudia, H. (2022). Designing transformation: Negotiating solar and green strategies for the sustainable densification of urban neighbourhoods. *Sustainability (Basel, Switzerland), 14*(6), 3438. https://doi.org/10.3390/su14063438

Frattolillo, A., Canale, L., Ficco, G., Mastino, C. C., & Dell'Isola, M. (2020). Potential for building facade-integrated solar thermal collectors in a highly urbanized context. *ENERGIES, 13*(21), 1JN-1JN. https://doi.org/10.3390/EN13215801

Fuldauer, L. I., Thacker, S., Haggis, R. A., Fuso-Nerini, F., Nicholls, R. J., & Hall, J. W. (2022). Targeting climate adaptation to safeguard and advance the Sustainable Development Goals. *Nature Communications, 13*(1), 3579. https://doi.org/10.1038/s41467-022-31202-w

Garshasbi, S., Haddad, S., Paolini, R., Santamouris, M., Papangelis, G., Dandou, A., Methymaki, G., Portalakis, P., & Tombrou, M. (2020). Urban mitigation and building adaptation to minimize the future cooling energy needs. *Solar Energy, 204*, 708–719. https://doi.org/10.1016/j.solener.2020.04.089

Grussa, Z. D., Andrews, D., Lowry, G., Newton, E. J., Yiakoumetti, K., Chalk, A., & Bush, D. (2019). A London residential retrofit case study: Evaluating passive mitigation methods of reducing risk to overheating through the use of solar shading combined with night-time ventilation. *Building Services Engineering Research and Technology, 40*(4), 389–408. https://doi.org/10.1177/0143624419840768

Gupta, R., & Gregg, M. (2020). Assessing the magnitude and likely causes of summertime overheating in modern flats in UK. *Energies, 13*. https://doi.org/10.3390/en13195202

Gupta, R., Kapsali, M., & Howard, A. (2018). Evaluating the influence of building fabric, services and occupant related factors on the actual performance of low energy social housing dwellings in UK. *Energy and Buildings, 174*, 548–562. https://doi.org/10.1016/j.enbuild.2018.06.057

Hitchin, P. (2014). Keeping the lights on. *Engineering & Technology, 9*(4), 36–39. https://doi.org/10.1049/et.2014.0401

IPCC. (2021). *IPCC AR6 WGIII*. https://www.ipcc.ch/report/ar6/wg3/downloads/report/IPCC_AR6_WGIII_Full_Report.pdf

Kennedy-Asser, A. T., Owen, G., Griffith, G. J., Andrews, O., Lo, Y. T. E., Mitchell, D. M., Jenkins, K., & Warren, R. F. (2022). Projected risks associated with heat stress in the UK Climate Projections (UKCP18). *Environmental Research Letters, 17*(3). https://doi.org/10.1088/1748-9326/ac541a

Lattimore, P. (2020, May 2020). Maintaining thermal comfort in a changing climate. *CIBSE Journal*. https://www.cibsejournal.com/technical/ensuring-thermal-comfort-in-a-warming-climate/

Lee, D. S., Jo, J. H., Koo, S. H., & Lee, B. Y. (2015). Development of climate indices using local weather data for shading design. *Sustainability (switzerland)*. https://doi.org/10.3390/su7021884

Lee, W. V., & Steemers, K. (2016). Exposure duration in overheating assessments: A retrofit modelling study. *Building Research & Information, 45*(1–2), 60–82. https://doi.org/10.1080/09613218.2017.1252614

Littlefair, P., Ortiz, J., & Bhaumik, C. D. (2010). A simulation of solar shading control on UK office energy use. *Building Research & Information, 38*(6), 638–646. https://doi.org/10.1080/09613218.2010.496556

Madge, G. (2021). *One billion face heat-stress risk from 2°C rise*. https://www.metoffice.gov.uk/about-us/press-office/news/weather-and-climate/2021/2c-rise-to-put-one-in-eight-of-global-population-at-heat-stress-risk

Makantasi, A., & Mavrogianni, A. (2016). Adaptation of London's social housing to climate change through retrofit: A holistic evaluation approach. *Advances in Building Energy Research, 10*(1), 99–124. https://doi.org/10.1080/17512549.2015.1040071

Mavrogianni, A., Pathan, A., Oikonomou, E., Biddulph, P., Symonds, P., & Davies, M. (2017). Inhabitant actions and summer overheating risk in London dwellings. *Building Research and Information, 45*(1–2), 119–142. https://doi.org/10.1080/09613218.2016.1208431

Mavrogianni, A., Paul, W., Davies, M., Phillip, B., & Oikonomou, E. (2012). Building characteristics as determinants of propensity to high indoor summer temperatures in London dwellings. *Building and Environment, 55*. https://doi.org/10.1016/J.BUILDENV.2011.12.003

Mavrogianni, A., Taylor, J., Davies, M., Thoua, C., & Kolm-Murray, J. (2015). Urban social housing resilience to excess summer heat. *Building Research and Information, 43*(3), 316–333. https://doi.org/10.1080/09613218.2015.991515

MHCLG. (2021). *The future homes standard: 2019 Consultation on changes to Part L (conservation of fuel and power) and Part F (ventilation) of the Building Regulations for new dwellings*. https://assets.publishing.service.gov.uk/government/uploads/system/uploads/attachment_data/file/956094/Government_response_to_Future_Homes_Standard_consultation.pdf

Nations, U. (2021). *The Sustainable Development Goals Report 2021*. United Nations. https://unstats.un.org/sdgs/report/2021/

Office, M. (2021). *UKCP18 headline findings report*. The Met Office.

Pye, S., Broad, O., Bataille, C., Brockway, P., Daly, H. E., Freeman, R., Gambhir, A., Geden, O., Rogan, F., Sanghvi, S., Tomei, J., Vorushylo, I., & Watson, J. (2021). *Modelling net-zero emissions energy systems requires a change in approach* (Vol. 21). https://doi.org/10.1080/14693062.2020.1824891

Rahif, R., Hamdy, M., Homaei, S., Zhang, C., Holzer, P., & Attia, S. (2022). Simulation-based framework to evaluate resistivity of cooling strategies in buildings against overheating impact of climate change. *Building and Environment, 208*, 108599. https://doi.org/10.1016/j.buildenv.2021.108599

Santamouris, M. (2016). Cooling the buildings—past, present and future. *Energy and Buildings, 128*, 617–638. https://doi.org/10.1016/J.ENBUILD.2016.07.034

Sharpe, T., McGill, G., Menon, R., & Farren, P. (2018). Building performance and end-user interaction in passive solar and low energy housing developments in Scotland. *Architectural Science Review, 61*(5), 280–291. https://doi.org/10.1080/00038628.2018.1502150

Su, M. A., Ngarambe, J., Santamouris, M., & Yun, G. Y. (2021). Empirical evidence on the impact of urban overheating on building cooling and heating energy consumption. *iScience, 24*(5), 102495–102495. https://doi.org/10.1016/j.isci.2021.102495

Tettey, U. Y. A., Dodoo, A., & Gustavsson, L. (2019). Design strategies and measures to minimise operation energy use for passive houses under different climate scenarios. *Energy Efficiency, 12*(1), 299–313. https://doi.org/10.1007/s12053-018-9719-4

Tillson, A. A., Oreszczyn, T., & Palmer, J. (2013). Assessing impacts of summertime overheating: Some adaptation strategies. *Building Research and Information, 41*(6), 652–661. https://doi.org/10.1080/09613218.2013.808864

UN. (2018). *2018 review of SDGs implementation: SDG 11* https://sdgs.un.org/sites/default/files/documents/20063197282018_background_notes_SDG_11_v3.pdf

Valitabar, M., GhaffarianHoseini, A., GhaffarianHoseini, A., & Attia, S. (2022). Advanced control strategy to maximize view and control discomforting glare: A complex adaptive façade. *Architectural Engineering and Design Management, 1–21*. https://doi.org/10.1080/17452007.2022.2032576

van Hooff, T., Blocken, B., Hensen, J. L. M., & Timmermans, H. J. P. (2014). On the predicted effectiveness of climate adaptation measures for residential buildings. *Building and Environment, 82*, 300–316. https://doi.org/10.1016/J.BUILDENV.2014.08.027

Vassiliades, C., Kalogirou, S., Michael, A., & Savvides, A. (2019). A roadmap for the integration of active solar systems into buildings. *Applied Sciences, 9*(12), 2462–2462. https://doi.org/10.3390/APP9122462

Waite, M., Cohen, E., Torbey, H., Piccirilli, M., Tian, Y., & Modi, V. (2017). Global trends in urban electricity demands for cooling and heating. *Energy, 127*, 786–802. https://doi.org/10.1016/J.ENERGY.2017.03.095

Widera, B. (2021). Comparative analysis of user comfort and thermal performance of six types of vernacular dwellings as the first step towards climate resilient, sustainable and bioclimatic architecture in western sub-Saharan Africa. *Renewable and Sustainable Energy Reviews, 140*, 110736. https://doi.org/10.1016/j.rser.2021.110736

Wright, A., & Venskunas, E. (2022). Effects of future climate change and adaptation measures on summer comfort of modern homes across the regions of the UK. *Energies, 15*(2). https://doi.org/10.3390/en15020512

Wong, L. (2017). A review of daylighting design and implementation in buildings *Renewable and Sustainable Energy Reviews*, 74959–968. https://doi.org/10.1016/j.rser.2017.03.061

# Built Environment Policy and Governance Innovations

# How Rwanda's Green Building Minimum Compliance System Can Help Achieve the Sustainable Development Goals

Ilija Gubić and Dheeraj Arrabothu

## Introduction

The adoption by United Nations member states of the 2030 Agenda for Sustainable Development, in September 2015, was a major milestone: for the first time, there was clear recognition from the international community of the need to focus on sustainable urbanization, with Sustainable Development Goal (SDG) 11, to "make cities and human settlements inclusive, safe, resilient and sustainable," dedicated to this aim (Gubić & Baloi, 2019). This goal includes a specific provision on sustainable buildings (SDG 11.c), which states that member states are to "support least developed countries, including through financial and technical assistance, in building sustainable and resilient buildings utilizing local material" (United Nations General Assembly, 2017). A number of targets under SDG 11, as well as other SDGs, are linked to functional, well-planned, and well-designed green buildings (Goubran, 2019; Goubran & Cucuzzella, 2019) (Table 1).

Implementing a green buildings and green cities strategy offers several environmental, economic, and social benefits that include the prevention of global warming and climate change, the reduction of carbon dioxide emissions and other pollutants, the protection of the ecosystem, the use of renewable natural resources, improved health, comfort, and well-being, the alleviation of poverty,

I. Gubić (✉)
Faculty of Architecture, University of Belgrade, Belgrade, Serbia
e-mail: 2020_41012@edu.arh.bg.ac.rs

D. Arrabothu
Global Green Growth Institute, Kigali, Rwanda
e-mail: arrabothu.dheeraj@gggi.org



T. Walker et al. (eds.), *The Role of Design, Construction, and Real Estate in Advancing the Sustainable Development Goals*, Sustainable Development Goals Series, https://doi.org/10.1007/978-3-031-28739-8_6

**Table 1** Linking green buildings with the SDG targets

| Number | Goal | Link |
|---|---|---|
| SDG 3 | Good Health & Wellbeing | Green building features, such as improved lighting, and better air quality, remain relevant to positively impacting health and wellbeing. Reducing emissions from buildings – particularly in cities – can reduce pollution and improve air quality, benefiting the health of city dwellers |
| SDG 7 | Affordable & Clean Energy | Green buildings promote the use of renewable energy, which can be less costly than fossil fuel alternatives. Renewable energies also have the added benefit of producing no carbon emissions, limiting their impact on the planet. Energy efficiency coupled with local renewable sources also improves energy security |
| SDG 8 | Decent Work & Economic Growth | As the demand for green buildings grows globally, so does the workforce required to deliver them. SDG 8 is, therefore, another goal that green buildings can significantly contribute to |
| SDG 9 | Industry, Innovation & Infrastructure | Green buildings must be designed in a way that ensures they are resilient and adaptable in the face of our changing global climate. This is critically important in developing countries, many of which will be particularly susceptible to the effects of climate change |
| SDG 11 | Sustainable Cities & Communities | Buildings are the foundations of cities, and green buildings are, therefore, key to their long-term sustainability. Whether it is homes, offices, schools, shops or green spaces – the built environment contributes to the make-up of communities, which must be sustainable to ensure a high quality of life for all |

(continued)

**Table 1** (continued)

| Number | Goal | Link |
|---|---|---|
| SDG 12 | Responsible Consumption & Production | The building industry has a major role to play in preventing waste through reduction, recycling, and reuse and the application of circular economy principles where resources are not wasted |
| SDG 13 | Climate Action | Buildings are responsible for over 30% of global greenhouse gas emissions and are therefore a major contributor to climate change. They also have the potential to combat it, offering one of the most cost-effective ways to do so through measures such as energy efficiency |
| SDG 15 | Life on Land | The materials that make up a building are key to determining its sustainability. Therefore, the construction industry and its supply chains have a major role to play in using responsibly sourced materials such as timber. Green building certification tools also recognize the need to reduce water use, the value of biodiversity and the importance of ensuring it is protected and incorporated into the space they build on both during and after construction – minimizing damage and designing ways to enhance biodiversity – such as through landscaping with local flora |

the improvement of economic growth, the rise of rental incomes, and a decrease in healthcare costs, among others (Ahn et al., 2013; Chan et al., 2017; Darko et al., 2017; Hikmat et al., 2009; Wang et al., 2018; Zhang et al., 2018). All of these factors support countries in achieving SDGs.

Globally, countries have implemented robust green building codes, rules, and regulations. As of September 2021, 80 countries had mandatory or voluntary energy codes for the building sector on the national or sub-national level, out of which 43 countries had mandatory codes on the national level for both residential and non-residential buildings (United Nations Environment Programme, 2021). The current extent of national and sub-national building energy codes worldwide shows that Sub-Saharan Africa and South and Central America have the least coverage of mandatory codes. In Africa, Morocco, and Tunisia have mandatory building energy codes that cover the entire building sector, while Egypt and South Africa have voluntary codes. Several countries in the region are in the process of developing building code standards, including Botswana, Burundi, Cameroon, the Ivory Coast, Ghana, Gambia, Kenya, Senegal, and Tanzania. In East Africa, Rwanda is the only country with a mandatory green building code that covers public (predominantly non-residential) buildings.

In 2019, the Rwanda Housing Authority, in collaboration with the Building Construction Authority of Singapore, the Global Green Growth Institute, the Rwanda Green Building Organization, and other stakeholders, developed the Green Building Minimum Compliance System (Arrabothu & Birungi, 2019). This system was approved by the Rwandan cabinet in April 2019 through a ministerial order determining the Urban Planning and Building Regulations and is an Annex-3 to the Rwanda Building Code of 2019 (Gubić et al., 2021; Republic of Rwanda, 2019). Implementing the green building rating system contributes to building performance in environmental aspects, among others (Hikmat et al., 2009; Khan et al., 2021; Remizov et al., 2021).

Against this background, this chapter seeks to assess the intersection between the institutional responsibility in planning, designing, and managing green buildings, the corresponding sustainable development goals, and how they are to be achieved in Rwanda. Consequently, this chapter seeks to determine the policy responses to rapid urbanization in Rwandan cities.

We gathered information from several learning events such as workshops, training sessions, and assessments through stakeholder interviews. More precisely:

- We observed ten workshops organized as part of the various Green Building Minimum Compliance System dissemination programs conducted from 2019 to 2022 which targeted multiple building industry stakeholders. This helped us to better understand Rwandan policies and to understand the willingness of the private sector to implement the policies in question.
- We actively supported the elaboration of the District Development Strategies for six of Rwanda's secondary cities from 2018 to 2019, where green building concepts were discussed with the local government and stakeholders through public sessions (Gubić & Baloi, 2020; Republic of Rwanda, 2018a). This helped us to identify specific issues related to green buildings in each district.
- We conducted a multi-country evaluation on green buildings to assess the transition of the building sector to more sustainable practices in different countries,

including Rwanda. The assessment looks at the regulatory environment, technical capacity, and the supply chain as some of the fundamental determinants of the pace of the transformation of the building sector. The assessment was focused on new buildings as well as the upgrade of existing buildings. This assessment was mostly conducted through stakeholder interviews representing the government, the private sector, and academia.

- We performed a document review to understand recent developments in policy measures introduced to further support green building implementation in Africa, and more specifically in Rwanda. The documents reviewed included academic papers and articles, as well as documents and reports prepared by the United Nations agencies and international organizations such as Global Green Growth Institute, International Energy Agency, World Green Building Council, and others working on green building standards.

## Theoretical Context

Many people might look at a building and see only an inanimate structure, but one needs to look at buildings and see both the structure and the process through which the building is created—as an opportunity not only to save energy and water and reduce carbon emissions but also to educate, create jobs, strengthen communities, improve health and well-being, and much more. Green building is a catalyst for addressing some of the most pressing issues in the world today.

Globally, the tangible and intangible benefits of green building vary depending on regional climate, building type, usage, occupants, number of hours the building is in operation, baselines used to validate savings, green building compliance ratings achieved by the building, and the country context. Generally, green buildings comprise the following dimensions:

- Energy efficiency and water conservation. Optimizing energy and water resources not only reduces the use of natural resources but also reduces water and energy expenditures.
- Improved indoor air quality. Green buildings reduce the need for air conditioning by maximizing natural ventilation.
- Reduced carbon footprint. By producing less waste and reducing the release of harmful gases, green buildings can lower their carbon footprint.

- Promotion of sustainable construction materials. Green buildings promote the adoption of building designs, construction practices, and materials that are environmentally conscious.

The built environment has a vital role to play in providing solutions to the global climate challenge. Green city development is uniquely positioned to address numerous sustainability issues, ranging from greener buildings to electric public transport. Globally, the building and construction sector accounted for 36% of final energy use and 37% of energy and process-related carbon dioxide ($CO_2$) emissions in 2020 (United Nations Environment Programme, 2021). In Africa, the building sector accounted for 61% of final energy use and 32% of energy-related carbon dioxide ($CO_2$) emissions in 2018 (United Nations Environment Programme, 2020). Recent estimates indicate that Rwanda's building sector, though small in terms of emissions at present at just under 1 $MtCO_2e$ in 2012, is set to grow to over 6 $MtCO_2e$ by 2050 under a Business as usual scenario (Global Green Growth Institute, 2019). This rise will be the result of an increase in the number of buildings and their associated energy consumption in the form of lighting, air conditioning, and electronic appliances.

Almost 60% of the population of the world will live in urban areas by 2030. Buildings are the foundation of cities, and green buildings are, therefore, the key to long-term sustainability. Whether it is homes, offices, schools, or shops—the built environment contributes to the make-up of communities which must be made sustainable to ensure a high quality of life for all. Hence, it is paramount for countries to promote green buildings in order to achieve SDG targets.

## Results and Discussion

Urbanization in Rwanda occurred only after a reconciliation process began after the Genocide against the Tutsi in 1994, with the ascent of a new government (Nduwayezu et al., 2021). The urbanization process was followed by Vision 2020, Vision 2050, the Economic Development and Poverty Reduction Strategy (I & II), the National Strategy for Transformation (I), the National Roadmap for Green Secondary Cities Development, master plans, and other new plans and policies related to green cities and green buildings (Baffoe, et al., 2020; Gubić & Baloi, 2019; Gubić et al., 2021).

In 2019, the Rwanda Housing Authority, in collaboration with the Building Construction Authority of Singapore, the Global Green Growth Institute, the Rwanda Green Building Organization, and other stakeholders developed the

Green Building Minimum Compliance System (Republic of Rwanda, 2019). This system was approved by the Rwandan cabinet in April 2019, through a ministerial order determining the Urban Planning and Building Regulations and is an Annex-3 to the Rwanda Building Code 2019 (Arrabothu & Birungi, 2019). Implementing the green building rating system contributes to the environmental performance of buildings, among other aspects (Khan et al., 2021; Remizov et al., 2021; Rouzbeh et al., 2017).

Implementing green buildings offers several environmental, economic, and social benefits to the construction industry including the prevention of global warming and climate change, minimizing carbon dioxide emissions and other pollutants, the protection of the ecosystem, the use of renewable natural resources, improved health, comfort, and well-being, the alleviation of poverty, improved economic growth, a rise in rental income, a reduction in healthcare costs, and others (Hikmat et al., 2009). The green building indicators in Rwanda address the minimum green features any building should have, such as appropriate orientations for daylighting, natural ventilation, rainwater harvesting, efficient plumbing fixtures, low-impact refrigerants, greenery protection, non-toxic paints, etc. These features can be applied to new Category 4 and 5 public buildings, such as health facilities, commercial buildings, educational buildings, cultural buildings, and others (Republic of Rwanda, 2019). The Green Building Minimum Compliance System in Rwanda also promotes the use of bricks as an energy-efficient building material that reduces heat ingress into a space by using advanced bonding techniques, such as the row-lock/rat-trap bond, thereby reducing the need for air conditioning systems and providing comfort to occupants (Arrabothu & Birungi, 2019; Republic of Rwanda, 2019; Photo 1) (Fig. 1).

## Linking Green Buildings' Minimum Compliance Point-Based System with SDG Indicators

The Rwanda Green Building Minimum Compliance System is comprised of 5 modules or focus areas targeting:

- Energy efficiency. This module focuses on the approach that can be used in the building orientation, design, material, and equipment selection to optimize the energy performance in building.
- Water efficiency. This module focuses on rainwater harvesting, selection of water-efficient fittings, waste water treatment, and other features that would reduce the use of potable water during building operation.

**Fig. 1** Hotel "Retreat" in Kigali, Rwanda, designed by ASA Studio (© Dheeraj Arrabothu, 2018)

- Environmental protection. This category focuses on the design, practice, and selection of materials and resources that would reduce the environmental impacts of built structures.
- Indoor environmental quality. This category focuses on the design strategies that would enhance the indoor environmental quality which includes fresh outdoor air provision, thermal comfort, noise reduction, and selection of non-toxic paints in buildings.
- Innovation and other green features. This category focuses on the adoption of green practices and new technologies that are innovative and have potential environmental benefits.

There are 29 green building indicators cutting across five modules which are defined in the document and are weighted for a total of 190 points. Each green building indicator is allocated points based on the relative importance of its contribution to green building goals. Points are allocated based on environmental impact, efforts required for implementation, and the costs associated with implementation.

Table 2 shows that there are 11 green building indicators defined under Module 1—Energy efficiency, which focuses on both passive design and active energy efficiency measures. These indicators are directly linked to the SDG target under

**Table 2** Linking the point-based Green Building Minimum Compliance System's module on energy efficiency with the SDG indicators

| S. No | Indicator | Optional/Mandatory | Points Allocation |
|---|---|---|---|
| 1.1 | Building Envelope – Façade Design Parameters | Optional | 25 |
| 1.2 | Ventilation | Optional | 25 |
| 1.3 | Daylighting | Optional | 17.5 |
| 1.4 | Artificial Lighting Efficiency | Mandatory | 5 |
| 1.5 | Enhanced Artificial Lighting Efficiency | Optional | 9 |
| 1.6 | Lifts and Escalators | Optional | 3 |
| 1.7 | Renewable Energy | Optional | 5 |
| 1.8 | Solar Hot Water System | Mandatory | 5 |
| 1.9 | Energy Metering | Optional | 2 |
| 1.10 | Air Conditioning System | Optional | 7.5 |
| 1.11 | Building Envelope – Air-conditioned Space | Optional | 8 |

*Source* Gubić and Arrabothu

Goal 7: Affordable and Clean Energy which aims to ensure access to affordable, reliable, sustainable, and modern energy for all. Some of the energy efficiency indicators are also linked to Goal 3 as good ventilation and lighting improve health and well-being, as well as Goal 13, since energy efficiency produces fewer emissions, helping to combat climate change.

Module 2—Water Efficiency has a total of 5 indicators, per Table 3, which focus on water conservation and management practices. These indicators are directly linked to Goal 15: Life on Land, as the indicators are aimed at conserving water resources through efficient water management practices that promote the concept of the 4 Rs: reduce, reuse, recycle, and refuse.

Module 3 of the Rwanda Green Building Minimum Compliance System—Environment Protection—has a total of 7 indicators (see Table 4) that are wide-ranging and focus on reducing embodied carbon emissions, biodiversity protection and enhancement, and waste management. These indicators are linked to Goal 12, where, by applying circular principles, resources are not wasted, as well as Goal 13, since lower embodied carbon emissions from buildings help to combat climate change, and Goal 15 as green building protects and preserves biodiversity and other essential resources. Indicators 3.4 and 3.5 also contribute

**Table 3** Linking the point-based Green Building Minimum Compliance System's module on water efficiency with the SDG indicators

| S. No | Indicator | Optional/Mandatory | Points Allocation |
|---|---|---|---|
| 2.1 | Rainwater Harvesting | Mandatory | 4 |
| 2.2 | Efficiency Plumbing Fixtures | Mandatory | 3 |
| 2.3 | Enhanced Efficient Plumbing Fixtures | Optional | 6 |
| 2.4 | Waste Water Treatment and Reuse | Mandatory Optional | 16 |
| 2.5 | Water Metering | Optional | 2 |

**Table 4** Linking the point-based Green Building Minimum Compliance System's module on environment protection with the SDG indicators

| S.No | Indicator | Optional/Mandatory | Points Allocation |
|---|---|---|---|
| 3.1 | Sustainable Concrete Usage | Optional | 5 |
| 3.2 | Greenery Protection | Optional | 10 |
| 3.3 | Environmental Friendly Practices | Optional | 2 |
| 3.4 | Low-impact Refrigerants: Zero Ozone Depletion Potential | Mandatory | 2 |
| 3.5 | Low-impact Refrigerants: Low Global Warming Potential | Optional | 4 |
| 3.6 | Segregation of Waste, Post-occupancy | Optional | 2 |
| 3.7 | Heat Island Mitigation | Optional | 3 |

to the Montreal Protocol[1] and the Kigali amendment to the Montreal protocol (United Nations, 2016).

Module 4—Indoor Environmental Quality (see Table 5) comprises 4 indicators focusing on promoting good indoor environmental quality and acoustics in buildings to improve occupant productivity, health, and well-being. These indicators are directly linked to Goal 3, as outdoor fresh air supply in air conditioned spaces and optimum air temperature improves indoor air quality and occupant comfort.

[1] Global environmental agreement adopted in 1987 that regulates the production and consumption of nearly 100 man-made chemicals referred to as ozone depleting substances (ODS). When released to the atmosphere, those chemicals damage the stratospheric ozone layer, Earth's protective shield that protects humans and the environment from harmful levels of ultraviolet radiation from the sun.

**Table 5** Linking the point-based Green Building Minimum Compliance System's module on indoor environmental quality with the SDG indicators

| S. No | Indicator | Optional/Mandatory | Points Allocation |
|---|---|---|---|
| 4.1 | Minimum Outdoor French Air Supply—Mechanically ventilated spaces | Mandatory | 4 |
| 4.2 | Thermal Comfort—Mechanically ventilated spaces | Mandatory | 2 |
| 4.3 | Noise Level | Mandatory | 2 |
| 4.4 | Low Volatile Organic Compound Paints & Adhesives | Optional | 2 |

**Table 6** Linking the point-based Green Building Minimum Compliance System's module on innovation with the SDG indicators

| S. No | Indicator | Optional/Mandatory | Points Allocation |
|---|---|---|---|
| 5.1 | Innovation | Optional | 10 |
| 5.2 | Universally Accessible Building | Mandatory | 9 |

Limiting noise levels within indoor spaces, and outdoors also promotes occupant comfort, while using fewer toxic paints and adhesives improves indoor air quality and reduces occupant exposure to toxic substances. Overall, the indicators are aligned and positively impact the health and well-being of occupants.

Module 5—Innovation and other Green Features (see Table 6) focuses on green building design and technologies that can spur innovation which are directly linked to Goal 9.

In addition to the above-mentioned direct links to the SDGs, the 29 green building indicators in the Green Buildings Minimum Compliance System also contribute to Goal 8 by creating new jobs and establishing new supply chains that can boost the economy.

Buildings contribute a quarter of global carbon emissions and account for a third of global energy consumption during construction and operation, and hence, considerable effort has gone into reducing energy consumption in buildings (Chung-Feng et al., 2016; Zepeda-Gill & Natarajan, 2020). The Third National Communication Report on Climate Change of the Rwandan Ministry of Environment estimates that the carbon dioxide emissions from buildings will increase by approximately 574% by 2050 compared to 2012 levels in a business as usual scenario (Arrabothu & Birungi, 2019; Republic of Rwanda, 2018b). However, the transition to green buildings comes with specific costs (Chan et al., 2018; Edwin et al., 2009; Fan et al., 2018; Häkkinen & Belloni, 2011; Qian et al., 2015).

Thus, the Government of Rwanda and other stakeholders are pursuing sustainability in the building and construction sector by recommending the use of energy-efficient practices and sustainably produced local construction materials, such as brick. They also recommend the use of energy-efficient kilns and alternative raw materials for firing instead of firewood, thereby reducing carbon dioxide emissions and protecting the environment during the brick production process (Republic of Rwanda, 2019). The updated Nationally Determined Contribution of the Republic of Rwanda, 2020, identifies efficient brick production as one of the key mitigation pathways in the energy sector (Republic of Rwanda, 2020). The use of locally-sourced raw materials for building material production, in this case, bricks, by reducing dependence on conventional materials such as concrete and steel, could potentially relieve pressure on the material supply chain, avoiding increased construction costs while reducing transport-related greenhouse gas emissions and providing opportunities for local economic development (Cheong & Storey, 2019).

The Made in Rwanda policy of 2017 by the Rwandan Ministry of Trade and Industry promotes the development of local construction materials, such as brick, in collaboration with the private sector, to reduce the trade deficit in construction materials (Gubić et al., 2021; Republic of Rwanda, 2017). The Assessing Rwanda's Affordable Housing Sector Report by the Centre for Affordable Housing Finance in Africa has noted that bricks are part of Rwanda's top 10 building material exports (Center for Affordable Housing Finance in Africa, 2019). This may be owing to Rwanda's abundant clay deposits, which are of excellent quality. There is also a massive demand in fast-growing cities for bricks (Swiss Resource Centre & Consultancies for Development, 2017).

## Promoting Climate Responsive Construction Material Production and Off-Farm Employment

The Promoting Climate Responsive Construction Material Production and Off-farm Employment in the Great Lakes Region (PROECCO) project of the Swiss Agency for Development and Cooperation (SDC), and implemented by SKAT Consulting Ltd, has showcased many innovations ranging from low-carbon brick manufacturing to modern brick construction systems, and by promoting semi-industrial brick manufacturing, which consumes 20–30% less energy in the production stage. The method uses biowaste, such as sawdust and coffee husks, for firing bricks. It also makes use of an innovative brick kiln technique that is adapted to a local context. This technique can consume up to 75% less energy than traditional methods, thus significantly reducing carbon dioxide emissions (Swiss Resource Centre & Consultancies for Development, 2017). The PROECCO project has demonstrated that the brick walls built using the row-lock/rat-trap bond technique are 30% cheaper than conventional cement block walls and also help in insulation and thermal comfort. This lower cost of construction combined with modular designs results in affordable construction that embraces brick as a key construction material that can potentially meet the growing affordable housing needs in Rwanda (Swiss Resource Centre & Consultancies for Development, 2017).

The revised master plans for Kigali, and the newly developed for six secondary cities, emphasize the inclusive nature of the participatory approach of developing polycentric cities with iconic cultural and traditional values. The revised Kigali City Master Plan Zoning Regulations mandates green building requirements by encouraging the use of local construction materials, including bricks, and provides developers with incentives that permit creating additional gross floor area if a project demonstrates a sustainable building design technology and sustainable construction methods as per the Green Building Minimum Compliance System (City of Kigali, 2020). In addition, District Development Strategies 2018–2024 recommend cities and other settlements within the district use brick and other sustainable construction materials (Gubić et al., 2021).

The Government of Rwanda and its stakeholders organized additional outreach events to promote the use of sustainable construction materials, especially brick, including "Urban Walk" (Gubić & Baloi, 2019). "Urban Walk" was initiated by the Global Green Growth Institute, in cooperation with the Ministry of Infrastructure and the University of Rwanda. Participants of the planned and guided walk, move through different city neighborhoods, and, on their walk, become familiar with issues of sustainable development and SDGs. In addition, awareness and capacity-building programs on Green Building Minimum Compliance System were organized for government officials, the Rwanda Institute of Architects, the Institution of Engineers Rwanda, and other professionals, where the usage of various green building materials and strategies has been promoted and emphasized (Arrabothu & Birungi, 2019, 2020).

The Green Building Minimum Compliance System has defined the indicators and their requirements specific to Rwanda's climate and development context. These include: designing an efficient western façade of the building, maximizing natural ventilation to take advantage of ambient upland climatic conditions, mandating rain water harvesting (as Rwanda has prolonged dry season spells leading to physical and economic scarcity of water resources), segregating waste at source to divert the segregated waste for recycling purposes and reducing the pressure on landfill sites, and promoting universal accessibility in public buildings to ensure social inclusion and to leave no one behind.

## Conclusion

The Rwanda Green Building Minimum Compliance System is also supporting the Government in its bid to promote green buildings by adopting sustainable construction practices, along with increasing the operational efficiency of buildings, in order to mitigate the emissions from the buildings sector, and thereby meeting the Nationally Determined Contributions (NDC) targets. The Government, with support from partners involved in the construction sector, is encouraging the implementation of Green Building Minimum Compliance System through awareness, outreach, and capacity-building programs to ensure the environment, economic and social benefits of green buildings trickle down the construction value chain. The usage of locally produced materials also helps in local economic development that has the potential to not just meet the growing construction material demand, but also to create decent off-farm employment opportunities for Rwandans as witnessed during the construction process.

Incorporating and implementing green building standards within the national building codes can support the transformation of the building and construction sector in a developing country as it transitions to a low-carbon and climate-resilient construction industry. The Green Building Minimum Compliance Systems aims to achieve this transition in Rwanda, and it could also serve as an inspiration for several countries in the region to encourage them to incorporate green building standards within their national building codes.

Many buildings in Rwanda demonstrate several green building principles that projects can adopt across the region to meet and potentially exceed the Rwanda Green Building Minimum Compliance System standard. At the same time as meeting new building standards, green buildings can create timeless contemporary architecture through the use of bricks and ultimately contribute to the triple bottom line of people, planet, and prosperity.

## Bibliography

Ahn, H. Y., Pearce, R. A., Wang, Y., & Wang, G. (2013). Drivers and barriers of sustainable design and construction: The perception of green building experience. *International Journal of Sustainable Building Technology and Urban Development, 4*(1), 35–45. https://doi.org/10.1080/2093761x.2012.759887

Arrabothu, D., & Birungi P. (2019). *Awareness workshop on the newly approved Rwanda Green Building Minimum Compliance System for architects*. https://gggi.org/awareness-workshop-on-the-newly-approved-rwanda-green-building-minimum-compliance-system-workshop-for-architects/

Arrabothu, D., & Birungi P. (2020). *Engineers from IER trained on Rwanda Green Building Minimum Compliance System and Templates at the Continuous Professional Development Workshop*. https://gggi.org/engineers-trained-on-the-rwanda-green-building-minimum-compliance-system-and-templates-at-the-continuous-professional-development-workshop/

Baffoe, G., Malonza, J., Manirakiza, V., & Mugabe, L. (2020). Understanding the concept of neighbourhood in Kigali city, Rwanda. *Sustainability, 12*, 1555. https://doi.org/10.3390/su12041555

Center for Affordable Housing Finance in Africa (CAHF). (2019). CAHF 2019 housing finance in Africa yearbook. CAHF.

Chan, A. P. C., Darko, A., & Ameyaw, E. E. (2017). Strategies for promoting green building technologies adoption in the construction industry—An international study. *Sustainability, 9*(6), 969. https://doi.org/10.3390/su906096

Chan, C. P. A., Darko, A., Olanipekun, O. A., & Ameyaw, E. E. (2018). Critical barriers to green building technologies adoption in developing countries: The case of Ghana. *Journal of Cleaner Production, 172*, 1067–1079. https://doi.org/10.1016/j.jclepro.2017.10.235

Cheong, C., & Storey, D. (2019). GGGI Technical report no. 4. meeting global housing needs with low-carbon materials. https://gggi.org/site/assets/uploads/2020/10/meeting-global-housing-needs-with-low-carbon-materials.pdf

Chung-Feng, J. K., Chieh-Hung, L., & Ming-Wen, H. (2016). Analysis of intelligent green building policy and developing status in Taiwan. *Energy Policy, 95*, 291–303. https://doi.org/10.1016/j.enpol.2016.04.046

City of Kigali. (2020). Zoning regulations, Kigali Master Plan 2050. https://masterplan2020.kigalicity.gov.rw/portal/sharing/rest/content/items/b9e31b07d8684053b0f3eac8649e1213/data

Darko, A., Zhang, C., & Chan, P. C. A. (2017). Drivers for green building: A review of empirical studies. *Habitat International, 60*, 34–49. https://doi.org/10.1016/j.habitatint.2016.12.007

Edwin, H. W., Chan, H. W. E., Qian, K. Q., & Lam, T. I. P. (2009). The market for green building in developed Asian cities—the perspectives of building designers. *Energy Policy, 37*(8), 3061–3070. https://doi.org/10.1016/j.enpol.2009.03.057

Fan, K., Chan, E. H. W., & Chau, C. K. (2018). Costs and benefits of implementing green building economic incentives: Case study of a gross floor area concession scheme in Hong Kong. *Sustainability, 10*(8), 2814. https://doi.org/10.3390/su10082814

Global Green Growth Institute. (2019). *GGGI Annual Report 2019, Greening urban planning and growth.* Global Green Growth Institute. https://report.gggi.org/2019/wp-content/uploads/2020/05/20027_Annual_Report_2019-FINAL_v05_JM_SS8.pdf
Goubran, S. (2019). On the role of construction in achieving the SDGs. *Hapres Journal of Sustainability Research, 1*, e190020. https://doi.org/10.20900/jsr20190020
Goubran, S. & Cucuzzella, C. (2019). Integrating the Sustainable Development Goals in building projects. *Hapres Journal of Sustainability Research, 1*, e190010. https://doi.org/10.20900/jsr20190010
Gubić, I., & Baloi O. (2019). Implementing the new urban agenda in Rwanda: Nation-wide public space initiatives. *Urban Planning, 4*, 223–36. https://doi.org/10.17645/up.v4i2.2005
Gubić, I., & Baloi, O. (2020) Public open space initiatives for healthier cities in Rwanda. *The Journal of Public Space, 5*(2), 129–146. https://doi.org/10.32891/jps.v5i2.1287
Gubić, I., Arrabothu, D., Bugirimfura, J., Hasabamagara, L H., Isingizwe, I., Kagina, A., Karigirwa, A., Mugema, J., Mukiza, A., Nishimwe, R., Nhimiyimana, W., & Yuhi, A. (2021). Advocating for Green Building Minimum Compliance System in Rwanda: Using bricks to achieve sustainability. *Facta Universitatis Series: Architecture and Civil Engineering, 19*(1), 67-80.https://doi.org/10.2298/FUACE210922006G
Hikmat, H., Ali, H. H., & Al Nsairat, F. S. (2009). Developing a green building assessment tool for developing countries—Case of Jordan. *Building and Environment, 44*(5), 1053–1064. https://doi.org/10.1016/j.buildenv.2008.07.015
Häkkinen, T., & Belloni, K. (2011). Barriers and drivers for sustainable building. *Building Research & Information, 39*(3), 239–255. https://doi.org/10.1080/09613218.2011.561948
Khan, M. A., Wang, C. C., & Lee, C. L. (2021). A framework for developing green building rating tools based on Pakistan's local context. *Buildings, 11*(5), 202. https://doi.org/10.3390/buildings11050202
Nduwayezu, G., Manirakiza, V., Mugabe, L., & Malonza, J. M. (2021). Urban growth and land use/Land cover changes in the post-genocide period, Kigali, Rwanda. *Environment and Urbanization ASIA, 12*(1), S127–S146. https://doi.org/10.1177/0975425321997971
Qian, Q., Chan, E., & Khalid, A. (2015). Challenges in delivering green building projects: Unearthing the Transaction Costs (TCs). *Sustainability, 7*(4), 3615–3636. https://doi.org/10.3390/su7043615
Remizov, A., Tukaziban, A., Yelzhanova, Z., Junussova, T., & Karaca, F. (2021). Adoption of green building assessment systems to existing buildings under Kazakhstani conditions. *Buildings, 11*(8), 325. https://doi.org/10.3390/buildings11080325
Republic of Rwanda. (2017). Ministry of trade and industry, made in Rwanda Policy. https://rwandatrade.rw/media/2017%20minicom%20made%20in%20rwanda%20policy%20(1).pdf
Republic of Rwanda. (2018a). District Development Strategy 2018a–2024. Vision: Rwanda's Business and Tourism Gateway. Government of Rwanda. http://197.243.22.137/rubavu/fileadmin/templates/raports/DDS_RUBAVU_28_DECEMBER_2018_AND___APROVED_FINAL_PDF_COPY.pdf
Republic of Rwanda. (2018b). Third National Communications under the United Nations Framework Convention on Climate Change (U.N.F.C.C.C.). Government of Rwanda.

https://rema.gov.rw/fileadmin/templates/Documents/rema_doc/TNC/The%20Third%20National%20Communication%20Report%20on%20Climate%20Change,%20Sept%202018b.pdf
Republic of Rwanda. (2019). Rwanda Building Code–Version 2. Government of Rwanda: Kigali. http://197.243.22.137/rhanew/fileadmin/user_upload/documents/General_documents/Laws_And_Regulations/rwanda_building_code_2019.pdf
Republic of Rwanda. (2020). Updated nationally determined contribution. http://climateportal.rema.gov.rw/sites/default/files/rwanda_updated_ndc_may_2020_0.pdf
Rouzbeh, S. R., Khorrami, M., & Ghaemi, M. (2017). Developing an Iranian green building assessment tool using decision making methods and geographical information system: Case study in Mashhad city. *Renewable and Sustainable Energy Reviews, 67*, 324–340. https://doi.org/10.1016/j.rser.2016.09.004
Swiss Resource Centre and Consultancies for Development. (2017). Modern Brick Construction Systems, a catalogue of affordable solutions, Made in Rwanda. http://madeingreatlakes.com/wp-content/uploads/2017/07/cat-modern-brick-solutions-draft-third-edition-beta-version2.pdf
United Nations. (2016). Amendment to the Montreal Protocol on Substances that Deplete the Ozone Layer. United Nations.
United Nations Environment Programme. (2020). GlobalABC Regional Roadmap for Buildings and Construction in Africa 2020–2050. Towards a Zero-emission, Efficient and Resilient Buildings and Construction Sector. United Nations Environment Programme. https://iea.blob.core.windows.net/assets/d60403aa-6fae-4365-bcd5-8af5daf22193/GlobalABC_Roadmap_for_Buildings_and_Construction_in_Africa_2020-2050.pdf
United Nations Environment Programme. (2021). 2021 Global status report for buildings and construction: Towards a zero-emission, efficient and resilient buildings and construction sector. United Nations Environment Programme. https://globalabc.org/resources/publications/2021-global-status-report-buildings-and-construction
United Nations General Assembly. (2017). Work of the Statistical Commission pertaining to the 2030 Agenda for Sustainable Development. United Nations General Assembly. https://documents-dds-ny.un.org/doc/UNDOC/GEN/N17/207/63/PDF/N1720763.pdf?OpenElement
Wang, H., Chiang, P. C., Cai, Y., Li, C., Wang, X., Chen, T. L., Wei, S., & Huang, Q. (2018). Application of wall and insulation materials on green building: A review. *Sustainability, 10*(9), 3331. https://doi.org/10.3390/su10093331
World Green Building Council. (2021). WorldGBC beyond buildings report. World Green Building Council. https://worldgbc.s3.eu-west-2.amazonaws.com/wp-content/uploads/2022/08/10190455/Beyond-Buildings_reduced_compressed.pdf
Zepeda-Gill, C., & Natarajan, S. (2020). A review of "Green building" regulations, laws, and standards in Latin America. *Buildings, 10*(10), 188. https://doi.org/10.3390/buildings10100188
Zhang, Y., Kang, J., & Jin, H. (2018). A review of green building development in China from the perspective of energy saving. *Energies, 11*(2), 334. https://doi.org/10.3390/en11020334

# Determining an Adequate Population Density to Achieve Sustainable Development and Quality of Life

Ahmed Salem

## Introduction

Population density is a crucial indicator in Urban studies, through which the relationship between the population and the land they live in is determined. It is also one indicator that determines the absorptive capacity of urban areas. Density can coordinate the relationship between the population and the built-up land, alongside the coordination of the horizontal and vertical relationship with the city. It is also a reflection of the city's urban properties; Population density is measured by calculating the ratio of the population to the area of land on which they reside. Population density is inversely related to the rate of land consumption; the higher the density, the less land consumption, and vice versa. There is a large diversity in densities from one city to another. Urban density is defined as the number of residences per unit of an area existing in a particular urban zone "Population density is linked to urban density, but to determine the criteria and parameters for this correlation, another study is required."

The original version of this chapter was revised: The incorrect word 'Low Density' in Fig. 11 has been removed now. The correction to this chapter is available at https://doi.org/10.1007/978-3-031-28739-8_12

A. Salem (✉)
General Organization for Physical Planning, Cairo University, Cairo, Egypt
e-mail: ahmedhelmy1@yahoo.com



T. Walker et al. (eds.), *The Role of Design, Construction, and Real Estate in Advancing the Sustainable Development Goals*, Sustainable Development Goals Series, https://doi.org/10.1007/978-3-031-28739-8_4

The term helps us to better understand the complex organism of a city (Cantera, 2020). The value of the city's population density reflects many aspects, including purchasing power, land prices, topography, the extent of the city's centrality, government policies, and the economic and social level of the people. High density "exceeds 450/ha" (Tonkin, 2008) is associated with walkability and dynamics related to healthy cities, affecting how cities deal with epidemics. Therefore, this chapter will discuss the relationship between density and the urban elements.

Some of the elements that affect the determination of the population density of the city can be identified as follows: -

1. Land prices.
2. The culture and lifestyle of the community.
3. Population characteristics (average income—socioeconomic status).
4. The shape of the urban fabric.
5. Infrastructure networks.
6. The city's location is distinguished.

In the past decade, many terms related to density, such as Urban Sprawl and consolidation, Smart Growth, and others, have appeared, and we will discuss some of them within this chapter.

This chapter aims to achieve sustainable density at the city and neighborhood levels. The chapter can also determine the upper and lower limits of sustainable density. In addition, we can determine factors affecting density according to its impact on achieving sustainable development and arrange them according to the strength of its coefficient. The study will also reach the equation that estimates the density value of the city according to the value set of urban variables. In addition, the study will determine the properties of cities located in the assumed Sustainable Density range.

## Methods

The study initially leans on previous research conducted on the determination of variables relating to density, alongside research conducted on determining the minimum and maximum appropriate density to achieve sustainable urban development. In the practical study, we will redetermine the minimum and maximum density suitable for sustainable development using SPSS statistical analysis and discriminate analyses, through which an equation is created that will redetermine the range of sustainable density. This will depend on the coefficient variable and will arrange them according to their impact on density.

## Urban Density

Urban density is a concept used in city planning, urban studies, and related fields to describe the intensity of people, jobs, housing units, total floor area of buildings, or some other measure of human occupation, activity, and development across a defined unit of area (Hess, 2014). It also has numerous definitions and measurement methods that seek to quantify urban characteristics, such as the total built-up area, floor area ratio (FAR), and the number of dwelling units, people, or jobs within a given area. This study will consider two levels of urban density: city density, which measures the inhabitants across a built-up area of the city; and district density, which measures the inhabitants across a built-up residential area of the district.

## Urban Sprawl

Urban Sprawl is the rapid expansion of the geographic extent of cities and towns, often characterized by low-density residential housing and single-use zoning. Shlomo Angel's study of more than sixty-five American cities over fifty years indicates that all the world's cities are heading toward an Urban Sprawl (Angel et al., 2010). Urban sprawl occurs when cities grow haphazardly away from their centers over previously undeveloped land in low-density patterns (Berggren, 2017). Seto's study conducted in 2011 showed that across a sample of 292 cities (equally distributed across the globe and measuring $< 100{,}000$ km$^2$ in extent, over the last three decades (1970–2000), urban land expansion rates are higher or equal to urban population growth rates (Seto et al., 2011).

## Smart Growth

Smart growth can be considered the opposite of sprawling; smart growth is usually planned and centers around the concept of compact town centers built around high-quality transit stations such as tram or subway stops. Smart growth, more specifically, is contained in an established urban corridor (Berggren, 2017). Examples of smart growth towns are, Fall Creek Place in Indianapolis, Indiana, USA, and Baltimore, Maryland, USA which both have mixed-use residential and commercial spaces, with restaurants, shops, and services, and it has easy access to downtown. Population density is the number of persons living in any given area (World bank, 2020).

## Importance of Density

Population density is a critical indicator in urban studies, through which the relationship between the population and the land can determine the capacity of urban areas. The study of urban density helps us better understand a city as a complex organism and determines the city's lifestyle, urban activities, and traffic. Urban density reports the necessary amount of people needed for achieving efficient development and traffic fluidity and facilitates urban functions while at the same time affecting the mixture of land uses, the variety of typologies, and the vitality of public space.

## The Objectives of Determining the Proper Density

Planners aim to optimize the city's urban density to achieve efficient services, suitable public space, and good traffic. This improvement in urban compactness allows for stable and resilient urban development, which can support sustainable development and smart growth. City densities must remain within a sustainable range. If the density is too low, it must be increased, and if it is too high, it must be reduced (Angel et al., 2010). On the other hand, cities with limited capacity such as those located in (agricultural areas, slums, and limited land availability), may not get the advantages of urban densification assumed for developed cities.

## Urban Densities Classification

Urban population densities vary widely from city to city according to many factors like location and spatial characteristics. We can classify cities according to urban density; for example, many Asian cities have some of the highest densities, frequently over 10,000 people per square kilometer and sometimes over 20,000 people per square kilometer, as found in Mumbai and Hong Kong (Frey, 1999). European cities have lower densities, in the range of 3,000 to 6,000 people per square kilometer, examples include Barcelona, Paris, Berlin, and Athens (Lehmann, 2010). North America and Australia have a low-density city typology with an urban downtown core surrounded by extensive urban sprawl, ranging from around 1,000 to 2,500 people per square kilometer; examples include Los Angeles, Phoenix, Melbourne, and Perth (Richardson, 2020). Each of these classifications has its own environment, geographical, and urban characteristics that affect the determination of its density (Lehmann, 2016).

## Advantages and Disadvantages of High Density

Planners' opinions differ in determining the appropriate density of the city; some argue that low density is associated with expansive growth and high resource consumption, while dense areas are associated with mixed land use, which allows for a blending of residential, commercial, and institutional activities in a given area. Mixed land use usually reduces transport costs and encourages walking and cycling due to the short distances (Teller, 2021). In addition, it has high-density sub-centers and high-density forms of housing. Lehmann has argued that a denser, more compact city is a more sustainable city (Lehmann, 2016). Roaf notes that high density is the inevitable urban future, with the need for protecting and conserving the open countryside. In addition, high density is linked to shortening travel times and the length of infrastructure networks (Roaf et al., 2009). Moreover, high density is also linked to cycling and public transport that save energy and reduce carbon emissions (Angel, 2012). At the same time, the disadvantage of high density is traffic congestion, infrastructure overload, overcrowding, air pollution, severe health hazards, lack of public and green space, and environmental degradation (Burgess, 2000). In addition, the compact city model produces high levels of noise pollution due to the proximity between dwellings and transport lines; we can note that in dense areas are a corresponding reduction in access to renewable energy, sunlight, and wind. Too much physical closeness can reduce daylight in buildings and limit access to solar energy.

Some planners see how the benefits of dense areas are limited and vary according to climate, land use type, culture, and latitude (Edwards et al., 2014). At the neighborhood level, the proximity of adjoining residential buildings creates problems of overlooking and noise. Smaller dwellings lack internal space between other houses, which results in a lack of privacy. However, people may accept the lack of privacy as a trade-off against other considerations (Dave, 2010). Figure 1 shows the variables that affect on density range, that this study had collected relying on literature.

## Effect of Density on Urban Elements

The density affects many urban elements with varying degrees of significance, but the effects of densification on housing affordability are complex. Some studies highlight that housing prices may become affordable when density increases due to the reduced size of housing units in compact development (short roads, efficient infrastructures) (Teller, 2021). Urban density is more attractive for companies

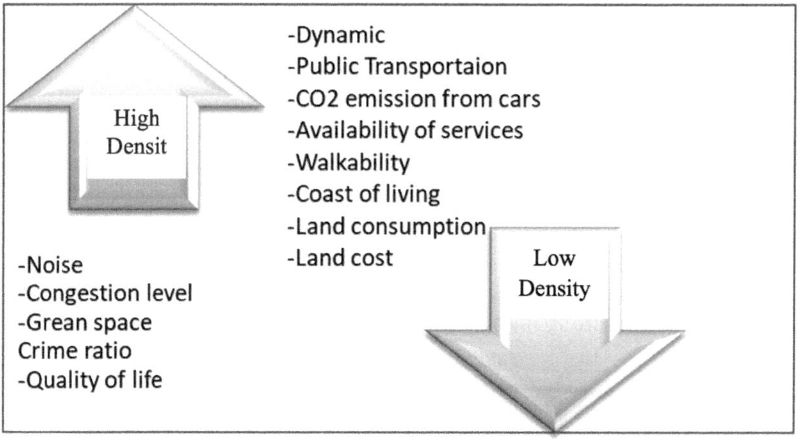


**Fig. 1** The variables affecting density range (author-originated)

through agglomeration effects in the service, business, and industrial sectors, better location with regard to existing centralities, and greater adherence to public life in the streets of residential areas (Montgomery, 1998).

## Population Density and Urban New Concepts

Urban density is associated with new urban concepts like dynamic, livable, healthy, resilient, and sustainable. We will discuss some new urban concepts as follows: -Density and dynamic.

Studies indicate a correlation between dense areas and dynamic cities; densification is expected to promote social interactions and cultural diversity. It is also associated with a lower level of urban fragmentation. Some studies suggest that high-density areas have higher social safety and tolerance (Todes et al., 2018).

### Density and Sustainability

On the other hand, alternative studies (Pozzi, 2005; Tonkin, 2008; Angle et al., 2012), argue that dense areas are associated with a various environmental, economic, and social indicator, including lower environmental impact and safer and more dynamic urban districts. Many modern urban planners advocate higher densities because of the widely held theory that cities operate more efficiently when residents live in denser urban surroundings. The sustainable density proposition should be moderate, neither too high nor too low, but "just right." Cities can achieve sustainability if its density is sufficient to provide a public transport network, which should not be less than 150 per ha or 65 people per ha. The city must be qualified to accommodate the above-average density to be considered sustainable. Determining the density value varies from one city to another according to the infrastructure characteristics, the availability of safety and security standards, and the distribution of services. Each group of cities can have a range of acceptable density that works to activate its positive role and achieve sustainability. Sustainability density (environment and economy) is closely associated with high medium people density (Bertaud, 2017, December 17). However, quality of life for humans is associated with a moderate density as it relates to the quality of the standard of living, whereas infeasibility of providing services is associated with low density.[1]

## Density and Resilience

High density makes urban systems more robust through dense area principles that improve and foster systems to be more resilient; density is one of the critical elements that affect the level of carbon dioxide and the rise in temperatures and, consequently, climate change. Many studies discuss the relationship between density and carbon dioxide, and all indicate a correlation between a decrease in density and an increase in carbon dioxide. Newman's study found that forty-six dense cities in 1990 were associated with shorter travel distances and lower energy expenditures on transport than lower density ones (Newman et al., 1999). In another study, Glaeser and Kahn studied carbon dioxide emissions from all transport modes in major cities in China and the United States in 2001; they found that average annual $CO_2$ emissions from transport in the U.S. cities were 56 times those in Chinese cities: 12.8 compared to 0.27 tons per household (Zheng et al., 2011). The average population densities in Chinese cities are 162 persons per ha compared to 23 in the U.S. cities. Figure 2 shows the average amount of $CO_2$ emissions per capita from all sources, and the average densities in cities with 100,000 people or more in 145 countries. We notice a direct relationship of close to 0.7 between the output of carbon dioxide and the decrease in density.

[1] We will explain high medium low density in detail later.

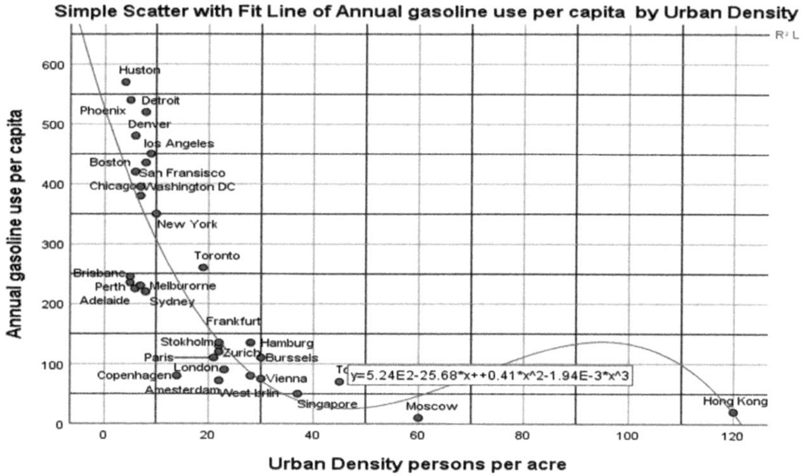


**Fig. 2** Relation between urban density and energy use (author-originated based on Newman, 1999)

## Adequate Density and a Healthy City

Many studies argue that high population density achieves a healthy life for its residents, as high density provides activity and dynamic for the city, interaction, and social relations between its inhabitants, and makes cities fit for pedestrians (Angel, 2012; Yu et al., 2022). This can be achieved through the availability of services, the economic feasibility of public transport, and commercial activities A British study (IBI Group, 2017) of 400 thousand people in twenty-two cities found that the population density to achieve a healthy city with good quality of living, available green spaces, parks, and pedestrian paths, should not be less than 65 people/feddan (Tonkin, 2008). The density required to reach certain benchmarks, such as access to public transport or livability, varies spatially and economically, for example for certain Arab cities a suitable density maybe 100 people/feddan, however Amsterdam, Netherlands density 22 people/feddan (Statista, 2020). Figure 3 shows the density frequency in Egyptian cities and its natural distribution curve and the mean urban density found the most frequent cities is between 70–90. As many Egyptian cities are in the agricultural areas and the hot, dry desert areas, the medium and high density is the nature of Egyptian

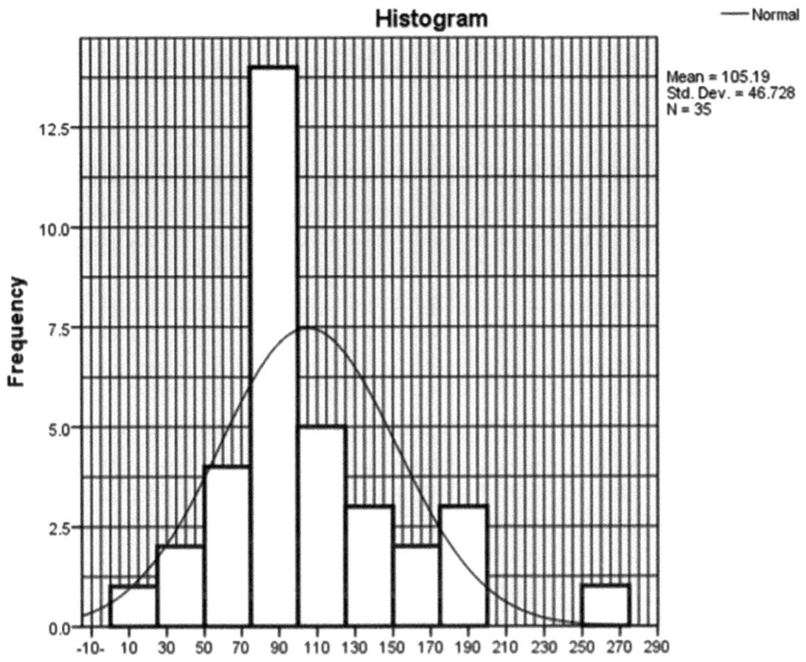


**Fig. 3** Frequencies of urban density for Egyptian cities (author-organized depending on CAPMAS data (CAPMAS, 2018))

cities. Taking into account that agricultural land in Egypt is scarce, it is forbidden to build on agricultural land and building heights are allowed, Therefore, the density of cities for them can range between (80–120) people per feddan, which is slightly higher than the average intensity shown by some previous studies.

The rapid spread of the Coronavirus varied between high- and low-density cities, due to Several variables we refer to a study applied to thirty-six cities showed no direct relationship between high density and the spread of COVID-19 Coronavirus. Another study (Duminy, 2021) noticed high-density cities such as Tokyo, Seoul, and Hong Kong, have succeeded in their prevention measures despite the high density of these cities (Adlakha et al., 2020). As people in high-density societies are more physically active (Pozzi, 2005), and their cities are more "walkable," which means there is a possibility of walking to shops, schools, and other services. It must be taken into account the strict measures on social distancing enforced by areas with high density, due to the speed of COVID-19 spread (Adlakha et al., 2020). Thus, leads to high-density cities have to be more strict restrictions than low density as the spread of epidemics in dense areas is more rapid.

## Tools and Methods for Determining Adequate Density

A United Nations study found relationships between ranges of adequate densities necessary to create viable thresholds for specific components of the urban environment. The study also tries to achieve an optimum balance. Based on this, an optimum gross density range is between 150 to 450 people per ha (approximately 150,00 and 450,00 people per km2, or 30 to 90 dwelling units per ha) (Tonkin, 2008). It also indicates that public transport is only viable[2] above 150 people per ha (Tuts et al., 2014). This study utilizes the values of indicators to determine the minimum and maximum sustainable density. The study will rely on Tonkin study (Tonkin, 2008), which found that 90 dwelling units per ha with an average family size of 4,9 (Pew Research Center, 2022), the density will be 360 people per ha*, and the maximum limit of new Egyptian cities is 120 people per feddan, 290 people per ha**. As the role of this study is to audit the density values by using discriminate analysis using the SPSS statistical analysis program, we will utilize the minimum and maximum values to determine adequate density to achieve sustainable development. The study will rely on two values, the minimum adequate density that achieves efficient public transportation (150 people/ha) (Tonkin, 2008), and the maximum adequate density. The UNDP's value of 90 dwelling units/ha (equal 360 people per ha) (Tuts et al., 2014), is close to the upper limits of the assumed density of the new Egyptian cities, at 290 people/ha.

In addition, this chapter studied 35 medium-sized Egyptian cities to determine the average density of Egyptian cities. The study found that the average density is 105 people/per feddan or 245 people per ha with a standard deviation of 46. Therefore, the upper and lower limits for the majority of city densities are "60, 150" people per feddan, or "143, 357" people per ha.

[2] Viable: it means more economical.

## Cities Studies and Density Inductors

Many studies dealt with ranking cities according to the extent they achieve new urban concepts, which use criteria like sustainability, resilience, and health. The criteria have been selected based on previous studies that have a significant effect on density and studies based on literature to determine the minimum and maximum density, which is almost 150 people/ha or 65 people feddan appropriate to provide public transport. The maximum limit is 300 people per ha or 130 people per feddan (UNDP & Tonkin, 2008), which can provide an urban quality environment. Based on the discrimination analysis using the SPSS program.

## Statistic and Discriminate Analysis

Discrimination analysis builds a predictive model for group membership. The model is composed of a discriminate function (or, for more than two groups, a set of discriminate functions) based on linear combinations of the predictor variables that provide the best discrimination between the groups. The functions are generated from a sample of cases for which group membership is known; The functions can then be applied to new cases with measurements for the predictor variables but with unknown group membership. Discriminant Analysis (IBM, 2022). Based on the variables identified from the literature and the values of the variables identified by ranks cities, according to these variables in world ranking studies, in addition to redefining and checking, the minimum and maximum density to achieve sustainable development the study prepared values of variables as follow:

## Case Study

The analysis relied on the study of 245 critical cities around the world (the capitals of countries, economic capitals, and megacities), representing continents, cities with high and low density, developed and rich countries, poor and developing countries. The study relied on five main variables drawn from the literature which directly impact the density; carbon dioxide emission, cost of living, crime index, noise, and congestion level. The Numbero website—the site for measuring indicators and analysis of investment and real estate development—provided the first three indicators of this data. The fourth indicator is Noise, measured by decibel units at different places in the city. The study relied on a sound print website as a reference, alongside the city council website (World Economic Forum, n.d.). The fifth indicator is the congestion levels in a place; the study relied on an application and study of traffic for cities of the world" cites (TomTom, 2021). The density data reference is Demographia World Urban Area 2022 and calculating the area through Google Earth and the city's population was taken from the census for each city (WPR, 2022). The density has been classified into three groups; the first is less than 150 inhabitants/ha, and as was mentioned previously, less than 150 is not suitable for public transportation; the second group is more than 150 and less than 300 inhabitants/ha, this is the most appropriate density to achieve sustainable development; the third group is more than 450 inhabitants/ha. Density (300 inhabitants/ha) represents the average of the two values: 150 and 450 people/ha, that suitable for people (Tuts et al., 2014), but 450 inhabitants/ha is too high for achieving sustainable development, so the study depends on (150,300) inhabitants/ha initially as limits of density for achieving sustainable development.

## The Relation Between District Density and City Density

The relation between city density and neighborhood's residential density. Can one of them be predicted in terms of the other? We studied thirty cities, for which we have density data on the city and district levels. Thus, we divided the density of the neighborhood (as it is the largest) by the density of the city and took the average, which is 2.5. Figure 4 shows the relation between district density and city. The study found that the districted density is almost 2.5-time city density. For dense districts, while included in the analysis, some low-density districts as Sant Andreu in Barcelona ratio is 0.65. So, we can consider the mean of two previous values (1.6) to determine the relationship between district density and city density. This is useful in predicting one of them in terms of the other.

## Analysis Using SPSS Program and Discriminate Analysis

We designated a code to each density cluster as this is a requirement for preparing a Discriminate Analysis; these codes indicate the ability to achieve sustainable

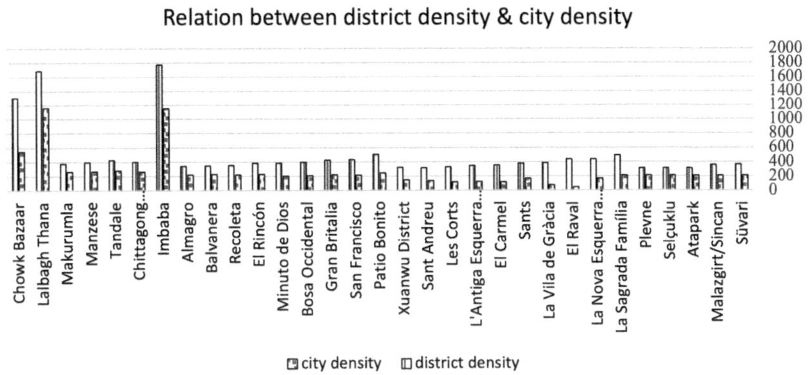


**Fig. 4** The relation between district density and city density (author-originated)

development. The cities of less than 150 people/ha took the number (1), the cities between 150- 300 people/ha took the number (2), and cities with the largest of 300 people/ha took the number (3). The analysis was carried out three times to test which equations are more representative of the classification of cities according to density (*):

- The first attempt at running analysis: is conducting a discriminate analysis through all variables, including density, to increase the effect of density on the equation.
- The second attempt is to enter all the variables without the density as the variable for classifying cities into groups. We can infer from the results the relative importance of each variable on the equation responsible for classifying cities according to density, which we will prepare through the SPSS Discriminate Analysis.
- The third attempt excludes the variables that have less impact on the equation, from previous attempts, while keeping the density variable to achieve the closest to the equation that matches the Initial classification.

**Table 1** Discriminate functions were used in the analysis (author-originated)

| Eigenvalues | | | | |
|---|---|---|---|---|
| Function | Eigenvalue | % of Variance | Cumulative % | Canonical correlation |
| 1 | 0.151[a] | 80.2 | 80.2 | 0.362 |
| 2 | 0.037[a] | 19.8 | 100.0 | 0.190 |

[a]First 2 canonical discriminant functions were used in the analysis

## Results

1. There are significant differences between the average of the variables (congestion level, cost of living,[3] and noise). This difference did not appear in the carbon dioxide emission variable or crime variables, as shown in Table 3 which shows the deference for each group for all variables in mean and stander divination. Figures 5, 6 and 7 also show the difference in the mean and median values of the three variables for each group.
2. Three attempts of discriminate analysis produced more than one equation; the closest to the initial classification is an equation that is related to the most influential variables, the ratio of matching between this equation and the results of classification is 80%, with the value of Eigenvalue 0.151, that expresses the performance of the equation.
3. The variables were also arranged according to the strength of its coefficient that impact on the density as discussed earlier. Table 1 refers to the influence of the equation including the density, and the second equation does not include the density, and we found that the most important variables affecting the density and fit the initial classification according to the second equation.

   As shown in Table 2, the results indicate; the cost-of-living level is 0.794, the level of traffic congestion is 0.789, and the noise is 0.295. This shows that the cost of living is the most important variable related to density, followed by traffic congestion and then noise. Figure 5 shows a Boxplot of cities according to their congestion level by density groups, we see a significant difference in the congestion level of cities depending on their density. Figure 6 shows a Boxplot of cities according to their cost of living by density groups, we see a significant difference in the cost of living of cities depending on their density. Figure 7 shows a Boxplot of cities according to their noise level by density

[3] Cost of living:—is a relative indicator of consumer goods prices, including groceries, restaurants, transportation and utilities. A Living Index of 120 it 20% more expensive than (New York) as a baseline.

**Table 2** Important variables that affect density (author-originated)

| | Function | |
|---|---|---|
| | 1 | 2 |
| cost of living | −0.794 | 0.032 |
| congestion level | 0.789 | −0.361 |
| noise level | 0.295 | 0.892 |

groups, we see a significant difference in the noise of cities depending on its density. Figure 8 shows the direct correlation between density and cost of living, it is clear from the figure that there is a direct relationship between them. Figure 9 shows the direct correlation between density and Noise, it is clear from the figure that there is a direct relationship between them. Figure 10 shows a direct correlation between density and congestion level, it is clear from the figure that there is a direct relationship between them.

4. New density Group classification

   Equations have been prepared based on the three attempts mentioned above, and we will rely on the third equation, which depends on the most influential variables—in addition to density variables—to redetermine the density clusters and its data are as follows: coefficients and constants. Table 4 indicates the final Classification Results of density groups, we can see that the first group has 236 cities within the initial classification with low density, 7 of these transferred from the first group to the second group in a new classification, while the second group (2) represents a medium density which falls within the assumed sustainability range. In the initial classification, there were 13 cities in the second Group, and seven of them were transferred from the first Group. The Third Group in the new classification deals with high density, that has 3 cities, so the number of cities in the sustainable density range is 13 out of a total of 245 cities.

4. Redefining sustainable development density range

   Discriminant equation redefines the limit of density groups and thus determines the extent of density that achieves sustainability, which is as follows;

- The density that achieves sustainability at the city level is 120 people/ha instead of 150 people/ha, and the upper limit is 240/per ha instead of 300 people/ha. Figure 11 Shows this density rang classification boundary, and shows a comparison between the Tonkin study and this study to determine density classification. This rang by feddan (50–100).

**Table 3** The different mean values of the variables by classes (author-originated)

| Density rank | | Mean | Std. Deviation | Valid N (listwise) | |
|---|---|---|---|---|---|
| | | | | Unweighted | Weighted |
| 1 | Congestion level | 26.1076 | 11.14558 | 158 | 158.000 |
| | $CO_2$ | 5257.0885 | 2588.23105 | 158 | 158.000 |
| | Cost of living | 62.1616 | 17.70182 | 158 | 158.000 |
| | Crime index | 42.2267 | 14.65929 | 158 | 158.000 |
| | Noise level | 72.3038 | 12.69621 | 158 | 158.000 |
| 2 | Congestion level | 37.9688 | 13.72421 | 32 | 32.000 |
| | $CO_2$ | 5444.6831 | 2072.69157 | 32 | 32.000 |
| | Cost of living | 44.5553 | 18.63069 | 32 | 32.000 |
| | Crime index | 48.1194 | 16.35540 | 32 | 32.000 |
| | Noise level | 74.0625 | 11.75540 | 32 | 32.000 |
| 3 | Congestion level | 47.7455 | 12.55837 | 11 | 11.000 |
| | $CO_2$ | 5846.3918 | 2430.51819 | 11 | 11.000 |
| | Cost of living | 28.2527 | 5.02411 | 11 | 11.000 |
| | Crime index | 50.3982 | 11.02572 | 11 | 11.000 |
| | Noise level | 85.3636 | 17.71029 | 11 | 11.000 |
| Total | Congestion level | 29.1801 | 13.16699 | 201 | 201.000 |
| | $CO_2$ | 5319.2047 | 2498.15530 | 201 | 201.000 |
| | Cost of living | 57.5029 | 19.80078 | 201 | 201.000 |
| | Crime index | 43.6120 | 14.95101 | 201 | 201.000 |
| | Noise level | 73.2985 | 13.13470 | 201 | 201.000 |

- The density that achieves sustainability at the neighborhood level is 192 people/ha, as the minimum and 384 people/ha as the maximum, "by multiply by city density by 1.6",That is, 80 people/feddan is a minimum, and 160 people/feddan is a maximum.

5. Properties of Sustainable Density range

This density range at the city level is 120–240 people/ha. If the density is more than 120 people/ha at the city level, the city at this level can provide public transport, the cost-of-living indicators will not exceed 28%

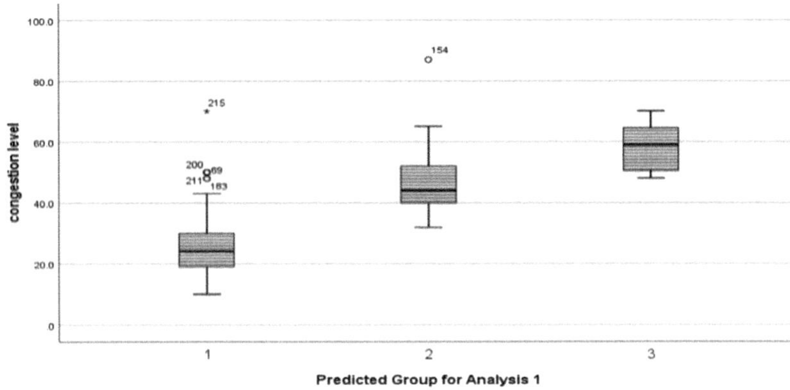


**Fig. 5** Boxplot of cities classes by congestion level (author-originated)

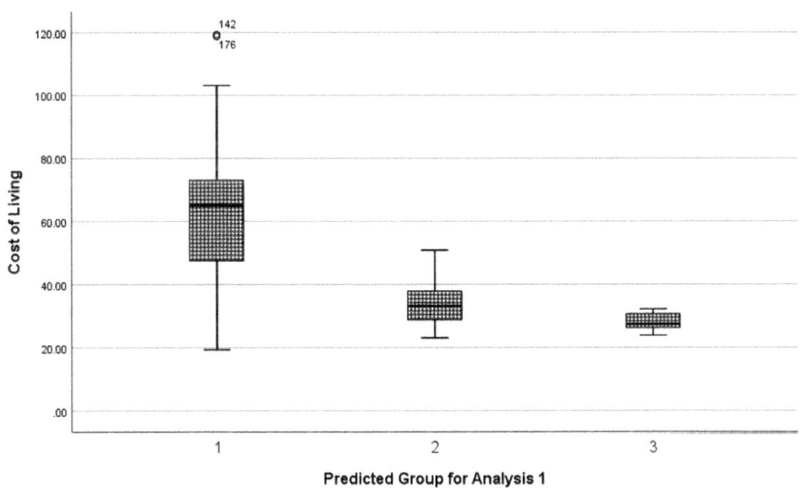


**Fig. 6** Boxplot of cities classes by the cost of living (author-originated)

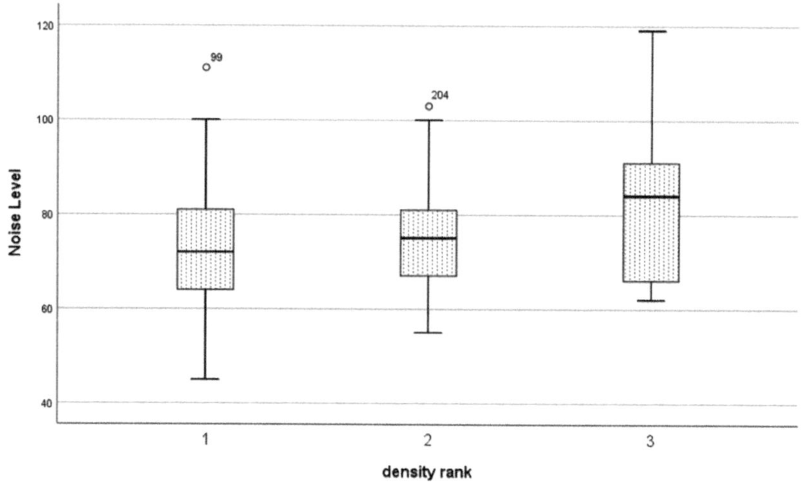


**Fig. 7** Boxplot of cities classes by noise level (author-originated)

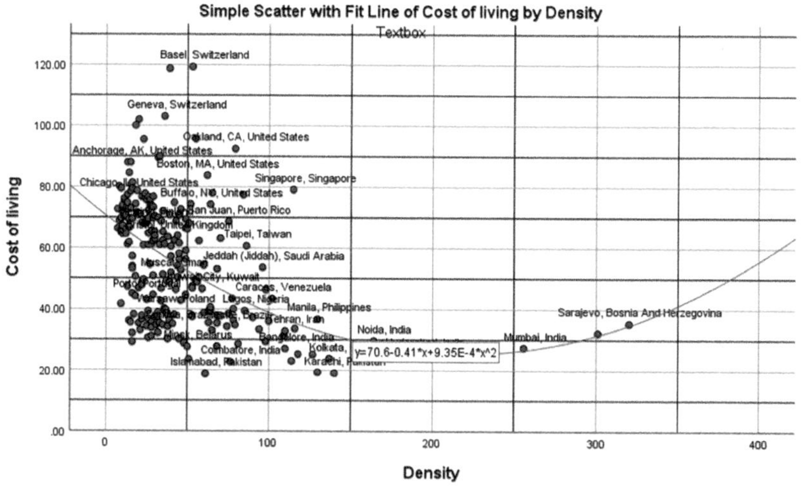


**Fig. 8** Correlation between density and cost of living (author-originated)

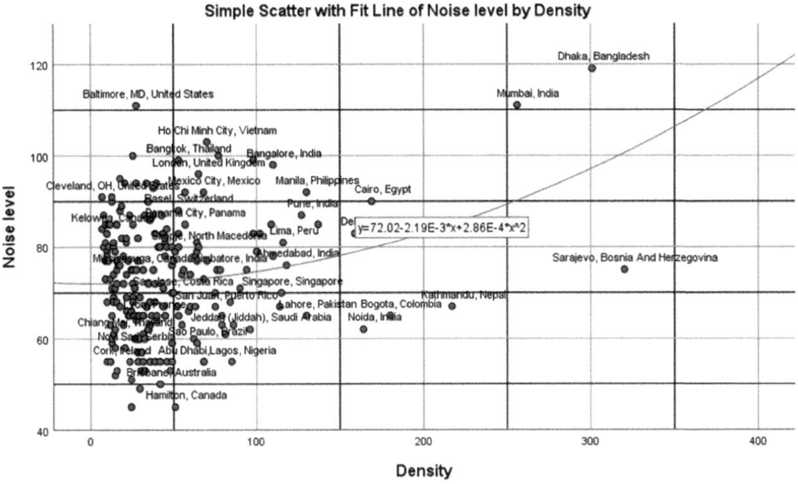


**Fig. 9** Correlation between density and noise (author-originated)

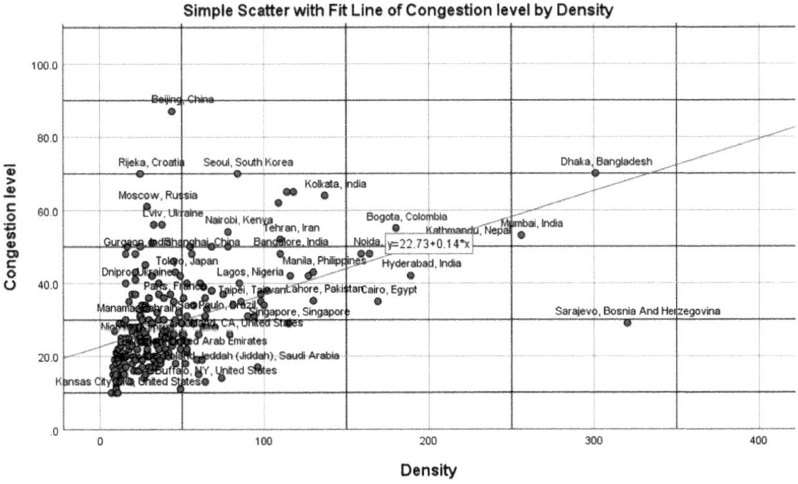


**Fig. 10** Correlation between density and congestion level (author-originated)

**Table 4** The final classification results of density groups (author-originated)

| Classification results[a,c] | | | | | | |
|---|---|---|---|---|---|---|
| Original | Count | 1 | 229 | 7 | 0 | 236 |
| | | 2 | 0 | 6 | 0 | 6 |
| | | 3 | 0 | 0 | 3 | 3 |
| | | Ungrouped cases | 2 | 0 | 0 | 2 |
| | % | 1 | 97.0 | 3.0 | 0.0 | 100.0 |
| | | 2 | 0.0 | 100.0 | 0.0 | 100.0 |
| | | 3 | 0.0 | 0.0 | 100.0 | 100.0 |
| | | Ungrouped cases | 100.0 | 0.0 | 0.0 | 100.0 |
| Cross-validated[b] | Count | 1 | 229 | 7 | 0 | 236 |
| | | 2 | 0 | 6 | 0 | 6 |
| | | 3 | 0 | 0 | 3 | 3 |
| | % | 1 | 97.0 | 3.0 | 0.0 | 100.0 |
| | | 2 | 0.0 | 100.0 | 0.0 | 100.0 |
| | | 3 | 0.0 | 0.0 | 100.0 | 100.0 |

[a]97.1% of original grouped cases correctly classified
[b]Cross validation is done only for those cases in the analysis. In cross validation, each case is classified by the functions derived from all cases other than that case
[c]97.1% of cross-validated grouped cases correctly classified

and annual gasoline use will not exceed 150 liters per capita. Moreover, if the density is less than 240 people/ha, the level of traffic congestion will not exceed 48%, meaning that the peak travel time does not exceed 48% on regular time of hours of the day. Also, if the density is less than 240, the average noise will not exceed 80 decibels in the city, as noise is one of the most important causes of disease and death (WHO Europe, 2020). The mentioned density limits support the achievement of sustainable development. Therefore, Thus, we have determined the density that achieves sustainable development, the most important factors affecting it, and the equation for calculating the city's maximum density based on its variables.

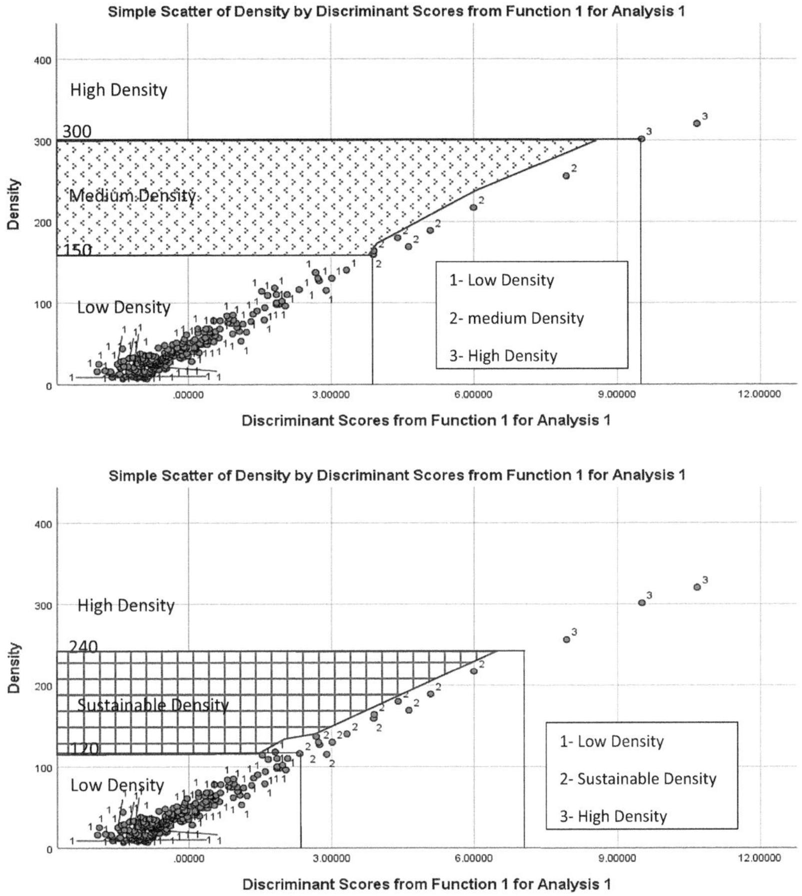


**Fig. 11** A comparison between the Tonkin study and this study to determine density classification and sustainable density (author-originated)

## Conclusion

Population density is a crucial indicator of Urban Studies, as density effect on noise, congestion of traffic, purchasing power, land prices, The density that achieves sustainable communities is associated with livable, walkable, agglomeration, dynamic, also it provides public transportation, and feasible access to services. The study determined sustainable density that ranged between 120–240 people/ha at the city level and between 192–384 at the neighborhood level. This is characterized by providing public transport, low cost of living (by 28% or less dense cities), the traffic congestion will not exceed 48%. And noise level less than 80 decibels. This study also determines factors affecting sustainable density and ordered them according to its impacts. The study also reaches the equation that estimates the density value of the city according to those factors. The character of sustainable density is different from one city to another according to its location, environment, availability of growth land and economic capacities. The appropriate density that achieves public transportation, traffic flow, livability, low level of noise, and cost enables us to achieve sustainable development.

## References


Adlakha, D., & Sallies, J. F. (2020, July 27). *Why urban density is good for health—even during a pandemic. Queen's Policy Engagement*. Retrieved December 19, 2022, from http://qpol.qub.ac.uk/why-urban-density-is-good-for-health-even-during-a-pandemic/

Angel, S. (2012). The sustainable densities proposition. Planet of cities. *Lincoln Institute of Land Policy*, 71–79.

Angel, S., Parent, J., Civco, D. L., & Blei, A. (2010). The persistent decline in urban densities: Global and historical evidence of sprawl. *Lincoln Institute of Land Policy*, Working Paper.

Berggren, C. C. (2017, June 20). *Urban density and Sustainability*. Smart Cities Dive. https://www.smartcitiesdive.com/ex/sustainablecitiescollective/urban-density-and-sustainability/241696/

Bertaud, A. (2017, December 17). The spatial distribution of population in 48 world cities. *Implications for Economies in Transition*. https://alainbertaud.com/wp-content/uploads/2013/06/Spatia_-Distribution_of_Pop_-50_-Cities.pdf

Burgess, R. (2000). The compact city debate: A global perspective. In M. Jenks & R. Burgess (Eds.) Compact cities: Sustainable urban forms for developing countries. *Spon Press*, 9–25.

Cantera, A. (2020, July 31). Multi-label urban density classification. urbanNext. Retrieved December 19, 2022, from https://urbannext.net/multi-label-urban-density/?fbclid=IwAR3jMogjt0VI8j2U2QM2-gSOlcZIYL30rI_9WewXrDDCOEqMM6Ly0p9QOHU

CAPMAS. (2018, July 1). Estimating the number of the Egyptian population in the sections of cities. https://www.capmas.gov.eg/Admin/PagesFiles/201892594224Untitled4.pdf

Dave, S. (2010). High urban densities in developing countries: A sustainable solution? *Built Environment, 36*(1), 9–27. https://doi.org/10.2148/benv.36.1.9

Duminy, J. (2021). Beyond growth and density: Recentering the demographic drivers of urban health and risk in the global south. *Urban Studies, 0*(0). https://doi.org/10.1177/00420980211014410

Edwards, B., & Hyett, P. (2014). *Rough guide to sustainability: A design primer* (4th ed.). RIBA Publishing.
Frey, H. (1999). Designing the city: Towards a more sustainable urban form. Taylor and Francis.
Hess, P. (2014). Density, urban. In A. C. Michalos (Eds.), *Encyclopedia of quality of life and well-being research,* 1554–1555. Springer. https://doi.org/10.1007/978-94-007-0753-5698
Habitat III. (2016). *The new urban agenda, UN,* https://habitat3.org/the-new-urban-agenda/
IBI Group. (2017, December 4). What is the healthiest density for a city. https://www.ibigroup.com/ibi-insights/healthiest-density-city
IBM. (2022). Discriminate analysis. https://www.ibm.com/docs/en/spss-statistics/25.0.0?topic=features-discriminant-analysis
Lehmann, S. (2016). Sustainable urbanism: Towards a framework for quality and optimal density? *Future Cities and Environment, 2*(1), 1–13.
Lehmann, S. (2010). *The principles of green urbanism.* Routledge.
Montgomery, J. (1998). Making a city: Urbanity, vitality and urban design. *Journal of Urban Design, 3*, 93–116. https://doi.org/10.1080/13574809808724418
Newman, P., & Kenworthy, J. (1999). *Sustainability and cities: Overcoming automobile dependence.* Island Press.
Pew Research Center. (2022, March 31). *Households smallest in Europe, biggest in Africa.* Pew Research Center. Retrieved December 18, 2022, from https://www.pewresearch.org/fact-tank/2020/03/31/with-billions-confined-to-their-homes-worldwide-which-living-arrangements-are-most-common/ft_2020-03-31_livingarrangements_01/
Pozzi, F., & Small, C. (2005). Analysis of urban land cover and population density in the United States. *Photogrammetric Engineering & Remote Sensing, 71*, 719–726. https://doi.org/10.14358/PERS.71.6.719
Richardson, H., & Gordon, P. (2020). Compactness or Sprawl: America's Future vs. The Present. *Atlanta*, the ACSP Conference. https://lusk.usc.edu/file/340/download?token=IntS7H4Q
Roaf, S., Crichton, D., & Nicol, F. (Eds.). (2009). Adapting buildings and cities for climate change (2nd ed.). Architectural Press.
Statista. (2020). Population density of Amsterdam from 2018 to 2020. *Statista.* Amsterdam population density 2020 | Statista.
Seto, K., Fragkias, M., Güneralp, B., & Reilly, M. K. (2011). A meta-analysis of global urban land expansion. *PLoS ONE, 6*(8), e23777. https://doi.org/10.1371/journal.pone.0023777
Teller, J. (2021). Regulating urban densification: What factors should be used? *Buildings & Cities, 2*(1), 302–317. https://doi.org/10.5334/bc.123
Todes, A., Weakley, D., & Harrison, P. (2018). Densifying Johannesburg: Context, policy and diversity. *Journal of Housing and the Built Environment, 33*(2), 281–299. https://doi.org/10.1007/s10901-017-9561-6
TomTom. (2021). Trafic study Ranking 2021. https://www.tomtom.com/traffic-index/ranking/
Tonkin, A. (2008). *Sustainable medium-density housing: A resource book.* Development Action Group.

Tuts, R., Rudd, A., & Swilling, M. (2014). *Leveraging density: Urban patterns for a green economy*. UN. Retrieved December 19, 2022, from https://unhabitat.org/leveraging-density-urban-patterns-for-a-green-economy

WHO. (2020). Promoting health and well-being throughout Europe. *WHO Regional Office for Europe*. https://apps.who.int/iris/rest/bitstreams/1365232/retrieve

World bank. (2020). data (worldbank.org). https://data.worldbank.org/indicator/EN.POP.DNST

World Economic Forum. (n.d.). *These are the cities with the worst noise pollution*. World Economic Forum. https://www.weforum.org/agenda/2017/03/these-are-the-cities-with-the-worst-noise-pollution/

World Population Review. (2022), *Amsterdam population*. Amsterdam Population 2022 (worldpopulationreview.com).

Yu, J., Gustafson, P., Tran, M., & Brauer, M. (2022, March 2). Assessing trade-offs and optimal ranges of density for life expectancy and 12 causes of mortality in Metro Vancouver, Canada, 1990–2016. *International Journal of Environmental Research and Public Health, 19*(5), 2900. https://doi.org/10.3390/ijerph19052900. PMID: 35270597.

Zheng, S., Wang, R., Glaeser, E. L., & Kahn, M. E. (2011). The greenness of China: Household carbon dioxide emissions and urban development. *Journal of Economic Geography, 11*(5), 761–792.

# Achieving the UN SDGs Through the Integration of Social Procurement in Construction Projects

David LePage and Emma Renaerts

## Introduction: Why Social Procurement?

Every purchase has an economic, environmental, cultural, and social impact, regardless of its size. Whether it be individual grocery choices, or government investment in a major infrastructure like a bridge, each choice entails many multipliers, ripple effects, and externalized consequences.

The power of the marketplace, when intentionally directed, has the potential to build not just economic value, but increase community capital, as shown in Fig. 1: healthy communities that are rich in human, social, cultural, physical, and economic capital.

A social value marketplace focused on achieving measurable outcomes like the UN Sustainable Development Goals (SDGs) will create communities that are inclusive, equitable, and diverse by offering training, employment, and social inclusion opportunities for every member of the community. In the last decade, as shown in Fig. 2, purchasers have increasingly considered the environmental impacts of their decisions.

In this chapter, we discuss the growing movement to add social value considerations to procurement decisions made in the construction industry, and we share case studies that demonstrate the power of procurement in construction to

D. LePage (✉) · E. Renaerts
Buy Social Canada CCC, Vancouver, Canada
e-mail: david@buysocialcanada.com

E. Renaerts
e-mail: emma@buysocialcanada.com



T. Walker et al. (eds.), *The Role of Design, Construction, and Real Estate in Advancing the Sustainable Development Goals*, Sustainable Development Goals Series, https://doi.org/10.1007/978-3-031-28739-8_7

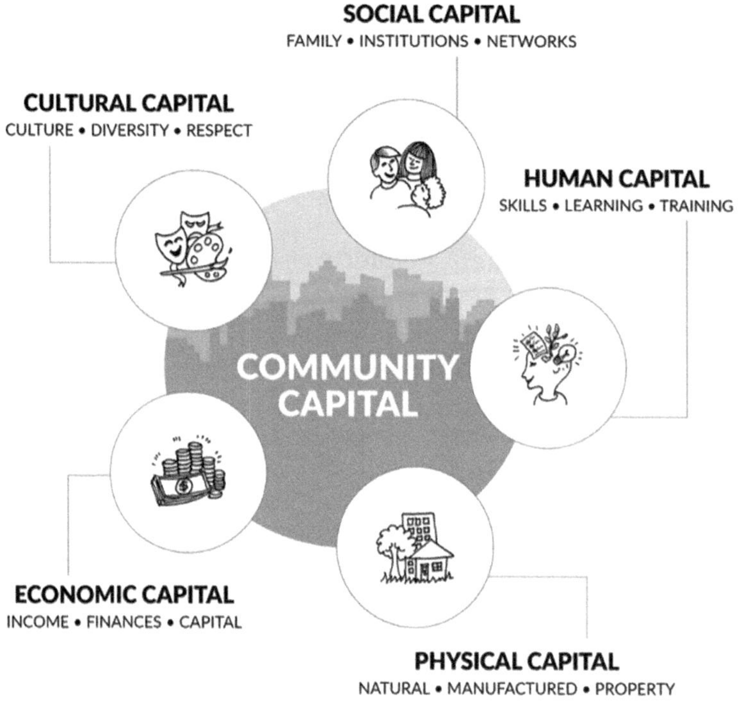


**Fig. 1** Community capital components (author-originated)

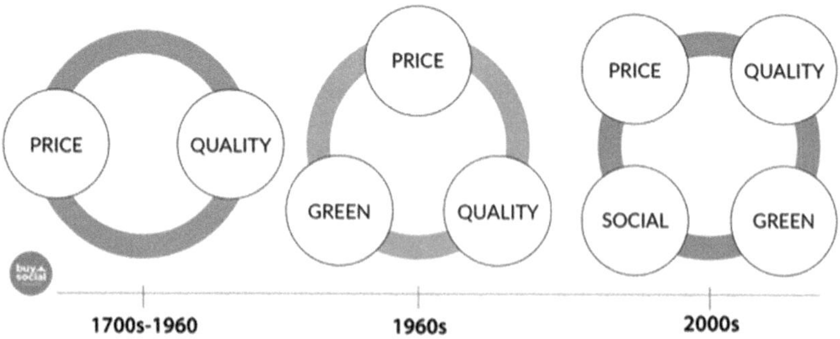


**Fig. 2** Procurement timeline (author-originated)

achieve multiple SDGs. Qualitative interviews with key figures from the organizations featured, along with research into relevant data and context, form the basis of the case studies presented below.

Social procurement, sometimes referred to as part of, or in terms of, sustainable procurement, asks how a purchase can help address and solve the complex socio-economic issues faced by a community. It seeks to consider where profits will go and how they will be used, who will be hired in the process, and who will provide the goods and services that go into the purchase. Simply put, it seeks to determine the community impact of the purchasing decision.

Figure 3 demonstrates how social procurement shifts marketplace value assumptions from considering purchasing as a purely economic transaction, trying to achieve the lowest price, to focusing on the power of the demand side of the market to drive positive social value outcomes. When we use social procurement to purchase goods, services, or choose a construction contractor, we're deliberately balancing the environmental impact, the social value outcomes, the product or service requirements, and the price.

The large scale of the purchasing power of governments includes billions of dollars of spending every year on construction projects and infrastructure investments. From school repairs to building new fire-houses, road replacements, or new bridges, each of these projects requires hiring labor and purchasing a myriad of goods and services.

Building on the strength of their purchasing power, governments have become an early adopters of social procurement initiatives:

> As the largest public buyer of goods and services, the Government of Canada can use its purchasing power for the greater good. We are using our purchasing power to contribute to socio-economic benefits for Canadians, increase competition in our

**Fig. 3** Social procurement process (author-originated)

procurements and foster innovation in Canada. (Public Services and Procurement Canada, 2020)

Across all levels of government, in several countries, including Canada, the United States, Scotland, Sweden, the United Kingdom, and Australia, policy is being designed and implemented to leverage existing purchasing or to use community benefit agreements to achieve social objectives and build community capital.

In general, over the past decade, we have witnessed the emergence of social procurement policies and initiatives across numerous entities as they adjust their historic purchasing criteria from lowest price to best value. Leveraging social value from existing buying offers an opportunity to solve persistent and complex social and environmental issues and work toward the accomplishment of the UN SDGs.

In Canada, Community Benefit Agreements (CBAs) between parties such as owners, developers, general contractors, and community members require certain social outcomes on a specific construction or infrastructure project. They are now the policy of the federal government, several provincial governments, and multiple municipalities from Victoria to Halifax, Calgary, Edmonton, Toronto, and others.

## Why Look to the Construction Industry to Achieve SDG's?

The economic purchasing power of the construction sector is significant: "In many ways construction is the backbone of the Canadian economy. It employs 1.4 million Canadians and accounts for 7.5 percent of Canada's GDP" (Canadian Construction Association, n.d.). Additionally, we know the "Canadian government will spend over $200 billion in the next ten years on infrastructure projects" (Office of the Parliamentary Budget Officer, 2022).

Imagine leveraging this enormous purchasing, hiring, and contracting power to not only build physical structures but to build community capital. This has already begun to happen. The construction industry is shifting from a business sector characterized as being driven by goals of "on budget and on time," and dominated by white and male employees, to an industry that integrates principles of social procurement and diversity in employment, and which is responsive to community needs. This shift means construction companies and purchasers can achieve multiple SDGs across the industry. The purpose of social procurement and community benefit models is to leverage the demand side of the construction industry market to achieve added social value. As demand increases for social value suppliers and a diversified workforce, more social value is created. This global paradigm shift toward social procurement and community benefits in construction is a clear path to achieving multiple Sustainable Development Goals. From ending poverty to reducing the impacts of climate change, the construction industry holds a unique set of keys to influence these outcomes.

Buy Social Canada is a social enterprise that advocates for and supports the design and implementation of social procurement policies and programs. In our work across a wide spectrum of projects, we have identified four key social value outcomes that can be achieved when social procurement is integrated into construction projects: (1) jobs, (2) training and apprenticeships, (3) social value in the supply chain, and (4) additional community development goals. Table 1 cross-references these outcomes with 12 SDGs which are most relevant to social procurement in construction. The construction industry addresses the SDGs when social procurement policies lead to collaboration with social value suppliers.

## Additional Benefits of Implementing Social Procurement in the Construction

This paradigm shift in procurement in the construction sector is being driven by multiple stakeholders who recognize the need for new processes and outcomes. A key element is the recognition that all levels of government and local communities around the world are experiencing severe and complex socio-economic issues such as unemployment, social exclusion, homelessness, youth disengagement, reconciliation with Indigenous populations, systemic racism, immigration, war, and poverty.

The construction industry also faces a serious worker shortage. The impending labor shortage is forcing a rethink of traditional construction labor market processes. The Canadian Construction Association reports that "about 21 percent of workers are set to retire over the next decade and the industry is struggling to attract the next generation of workers" (Canadian Construction Association, n.d).

Chandos Construction President Tim Coldwell adds: "Our industry has a huge problem. We are currently short 200,000 skilled workers, and 46% of our current labor force will retire in the next 10 years" (T. Coldwell, personal communication, July 15, 2022). Another General Contractor states:

> If we stick to the same old traditional approach we've used for 100 years, we won't get anywhere. The diversity piece is the big benefit in terms of increasing our supply

**Table 1** The UN SDGs and social value outcomes through construction (author-originated)

| SDG goal | Jobs | Training & apprenticeships | Social value supply chain | Community development |
|---|---|---|---|---|
| #1: No poverty | ✓ | ✓ | ✓ | |
| #2: Zero hunger | ✓ | | | |
| #3: Good health and well-being | ✓ | | ✓ | ✓ |
| #4: Quality education | | ✓ | | |
| #5: Gender equality | ✓ | ✓ | ✓ | |
| #8: Decent work and economic growth | ✓ | ✓ | ✓ | ✓ |
| #9: Industry, innovation and infrastructure | | | ✓ | ✓ |
| #10: Reduced inequalities | ✓ | ✓ | ✓ | |
| #11: sustainable cities and communities | | | | ✓ |
| #12: Responsible production and consumption | | | ✓ | |
| #13: Climate action | | | ✓ | ✓ |
| #17: Partnerships for the goals | | | ✓ | ✓ |

> chain… We need different approaches and different options, thoughts, and experiences. (Buy Social Canada, 2022, pp. 53–54)

Adding in social value considerations can help to address these labor market shortages by tapping into a pipeline of new workers from historically marginalized or barriered groups, and sourcing new sub-contractors who deliver on social value.

The industry is also grappling with its environmental impacts, recognizing the ways waste and building materials contribute to climate change:

> According to new research by construction blog Bimhow, the construction sector contributes to 23% of air pollution, 50% of climatic change, 40% of drinking water pollution, and 50% of landfill wastes. In separate research by the U.S. Green Building Council (USGBC), the construction industry accounts for 40% of worldwide energy usage, with estimations that by 2030 emissions from commercial buildings will grow by 1.8%. (GoContractor, 2017)

Practicing social procurement processes which prioritize suppliers who are mitigating environmental impacts is one way the construction industry can begin to address this trend.

## Social Value in the Supply Chain

Social enterprises are businesses that sell goods and services, embed a social, cultural, or environmental purpose into the business, and reinvest the majority of profits into their social mission. They provide a multitude of services to the construction industry supply chain, including staffing and labor, trailer cleaning, junk removal, printing and signage, catering and food trucks, security, couriers, and much more.[1]

As most sites have multiple sub-contractors and trades on each project, providing jobs, creating training and apprenticeships, and using social value suppliers is not just the role of the primary contractor. Owners, general contractors, sub-contractors, and any others involved on a project should consider every opportunity available onsite to amplify social impact.

If a social enterprise is not immediately available, purchasers can use a concentric circle process (shown in Fig. 4) to look for other social value suppliers. Depending on their goals, this could be an Indigenous, Black, or women-owned business, a co-operative, a B Corp, or a locally owned or operated business.

### Case Study: EMBERS, Supportive Employment, and Community Economic Development

EMBERS, the Eastside Movement for Business and Economic Renewal Society, is a social enterprise and registered community economic development charity

[1] In Canada, see the Buy Social Canada Directory of Certified Social Enterprises: https://www.buysocialcanada.com/directories/.

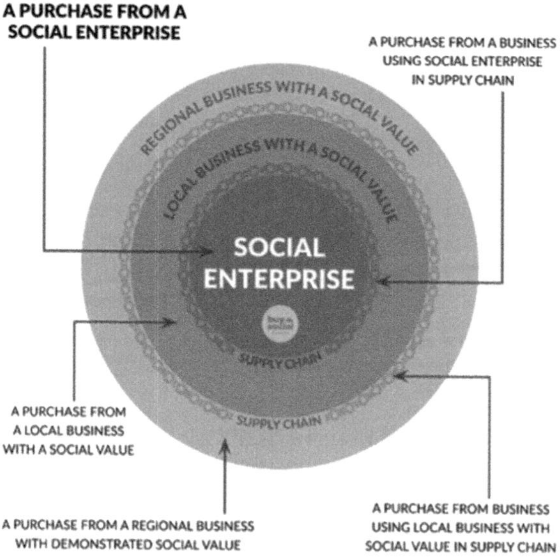


**Fig. 4** Concentric circle decision-making in the supply chain (author-originated)

located in Vancouver's Downtown Eastside, one of Canada's poorest neighborhoods. Since its founding in 2001, EMBERS has worked to create economic and employment opportunities for people living on low incomes (EMBERS Canada, n.d.). In 2008, to further this mission, they started EMBERS Staffing Solutions, a non-profit staffing agency, "placing all of [their] profit back into our workers through services like training, free equipment rental, and transportation" (EMBERS Staffing, n.d.). Their work has been recognized with many awards and accolades, including a 2021 Governor General's Innovation Award (Governor General's Innovation Awards, 2021). "Our focus is to work with people who are not necessarily conditioned to 'normal' 9–5 work or middle-class life, what is expected when you think of work. We're working with people who are learning how to get back and develop a sustainable income," CEO and Founder of EMBERS, Marcia Nozick, told Buy Social Canada in an interview. "Our social enterprise model can meet people where they're at" (M. Nozick, personal communication, June 15, 2022).

Construction companies have been valuable partners for EMBERS. Most EMBERS employees are placed in construction jobs, which are uniquely suited to support people looking to re-enter the workforce and start with what they're able to manage. Construction companies tend to use a lot of contingent labor, and most projects have different stages where various kinds of labor are needed. "As a result, you can work with that to mold something to fit the right person to the right job," says Nozick.

Construction is also a great workplace for EMBERS employees, Nozick shares, because it is easy to "ladder up" to a career over time. EMBERS has seen several people go from general work on sites, to eventually become site supervisors. One such employee is Neil (name changed to protect anonymity). Nozick shares that Neil had experienced severe trauma in his life. An immigrant to Canada, he lost his family in tragic circumstances and now struggled with anxiety and other mental health issues. For many years, he could not work at all. After several years of treatment, however, his counselor recommended he work with EMBERS. "He had the greatest attitude," says Nozick. Starting on site as a general laborer, Neil had a rocky beginning, sometimes showing up, sometimes not. One day, Nozick says, Neil realized, "actually work makes me feel better, work is my therapy." Gradually, he began to work more steadily and was able to take several training courses with support from EMBERS. He qualified as a hoist operator, and then a construction safety officer.

After three years of working with EMBERs, Neil's talents and dedication were recognized by a large construction company, who hired him full-time and is planning to put him through school to further his career development. He is now the head construction safety officer of his company, and in his role, he employs more EMBERS workers. He told Nozick that EMBERS saved his life.

Another EMBERS employee who was able to build a career in construction is Mike (name changed to protect anonymity). When Mike first started with EMBERS, he had just left treatment for a drug addiction and was homeless. No one would hire him until he was referred to EMBERS. "We had him working the very next day," Nozick says, "and Mike never looked back." Today, Mike is a site superintendent for a local construction company, and he has since re-married and bought his own apartment.

Every year, EMBERS employs 2,500 people with stories like Neil's and Mike's, and "every one of those people need[s] a job," says Nozick. The work they do achieves many Sustainable Development Goals, most notably: SDG 1, No Poverty; SDG 4, Quality Education; SDG 8, Decent work and Economic growth; and SDG 11, Sustainable Cities and Communities.

In addition to creating opportunities for many who are unemployed or underemployed and providing training and wrap-around support to them, EMBERS, as a social enterprise, also re-invests the majority of its profits back into the community. One key lever that enables EMBERS to achieve these outcomes is social procurement. "Social procurement opens up opportunities for us as a social enterprise to grow," says Nozick, "it really help[s] to build our foundation."

Nozick also notes that social procurement, for example, under the City of Vancouver's Community Benefits Agreement (CBA) Policy, "encourages patience on the part of the contractor," because they are willing to work with social value suppliers and create some space for employees with barriers, as part of their contract requirements on a project. "We wouldn't be successful without companies buying in," says Nozick, "and social procurement is one of those 'tying-together' pieces that helps communities connect. I think it's a really great strategy."

Construction companies can choose from many temporary labor services, but only EMBERS offers a full program of employee support (including training, insurance, good wages, and even encouragement) to help their workers to move on to full-time permanent work through collaborative programs with their business clients. Neil and Mike's stories serve to illustrate and corroborate the fact that the collective activities of EMBERS achieve at least a dozen of the UN SDGs every day, an effect that would be amplified if similar concepts were adopted globally.

In Canada alone, there are many other effective construction social enterprises seeking to create healthy communities and support people through employment. BUILD, a social enterprise in Winnipeg, Manitoba, trains and employs youth at risk in the building trades. Build Up Saskatoon offers steady employment, mentorship, and other social supports to their employees to aid them to achieve meaningful, long term employment in the construction industry. Building Up in Toronto works with persons facing barriers to enable them to enter the labor market through jobs in the construction industry. Impact Construction in St. John's, Newfoundland, has built their business model around supporting youth to gain skills and build a successful future.

The same trend is happening globally. Veterans in Construction in Melbourne, Australia, is "a Veteran owned and operated company […] committed to providing long-term sustainable employment opportunities to Veterans within the construction industry" (Veterans in Construction, n.d.). In London, Bounce Back trains and employs people leaving prison to enter construction jobs. And in Mexico, Echale a Tu Casa trains people in construction skills, while helping to build affordable homes for Mexico's poorest communities.

## Policy as a Driver for Change

Governments at all levels and in multiple countries see the opportunity and the need to leverage social value outcomes to meet their social goals and to address equity, social inclusion, poverty, climate change, and more, all of which align with the UN SDGs. As mentioned earlier, the city of Vancouver, British Columbia, has a mandatory Community Benefit Agreement (CBA) Policy which requires large, rezoned development projects to include the creation of specified local and social jobs and the inclusion of a diverse and social value supply chain (City of Vancouver, n.d.). Canada's Federal Ministry of Infrastructure has adopted a Community Employment Benefits program to promote increased employment opportunities "for a broader array of Canadians" (Government of Canada, n.d.). In Scotland, the Public Reform Act is used "to improve the economic, social or environmental wellbeing" of the communities in which construction projects occur (Scottish Government, n.d.).

As Government policy increasingly requires proponents to create social value from construction projects, we see that "the evidence of best value through procurement is still a nascent movement, with a need to address pre-existing perceptions through research and case study evidence"(Buy Social Canada, 2022). Some early criticism from industry sees the emerging policies as extortion for social value, a distraction, or merely an added cost: "This survey of CBAs points to the tantalizing potential of a good idea [...] there is much work to be done if CBAs are to truly achieve their promise"(Cardus, 2021).

Dr. Daniela Troje interviewed key actors in the Swedish construction sector and reviewed policy-in-practice literature. She concluded that while social procurement policies can mitigate issues connected to social exclusion, unemployment, and segregation, and that while the construction sector holds opportunities for social procurement, there is currently a misalignment among social procurement policies (Troje, 2021).

In their 2021 research on champions of social procurement in the Australian construction industry, Loosemore, et al. find that:

> There has been a recent proliferation of social procurement policies in Australia that target the construction industry. This is mirrored in many other countries, and the nascent research in this area shows that these policies are being implemented by an emerging group of largely undefined professionals who are often forced to create their own roles in institutional vacuums. (Loosemore, Keast, Barraket, et al., 2021)

Though the movement is still in its early stages, it already serves as proof that policy has the power to influence change and help achieve the UN SDGs in the construction industry.

## Case Study: Chandos Construction: Policy as a Competitive Advantage and Value for the Construction Sector

Chandos Construction is an exemplary business within the construction industry. Established in 1980, they were founded from the beginning with a vision to "be a business that balances both profits and people" (Chandos Construction, n.d. b). A company of over 600 employees spanning seven locations across Canada, they are the first and largest B Corp certified national technical builder in North America and are 100 percent employee-owned (Chandos Construction, n.d. b).

Chandos is working toward achieving many UN SDGs in their work. Early adopters of social procurement in the construction sector, they signed on as founders of the Buy Social Canada Pledge in 2020, committing to "shift at least five percent of [their] addressable spend to social enterprises, diverse-owned businesses, and other impact organizations by 2025" (Chandos Construction, n.d. c). In 2022, they announced an additional commitment to become Net Zero by 2040, working toward increased sustainability and reduced environmental impacts (Chandos Construction, n.d. a). President Tim Coldwell, states that the top three SDGs Chandos focus on through their work are SDG 8, Decent Work and Economic Growth; SDG 1, No Poverty; and SDG 4, Quality Education.

Through social procurement, Chandos uses its labor force to provide employment opportunities for local people from equity-deserving groups (Chandos Construction, n.d. c). Its aim is to:

- Expand diversity in local businesses.
- Reduce poverty and strain on the Canadian social system.
- Provide work, offer skills training, and pay fair wages to people from equity-seeking groups.
- Support businesses that are working toward the goals of the Paris climate agreement.

In a qualitative interview with Buy Social Canada, Coldwell shared the many ways that inclusive hiring can achieve the SDGs, and save taxpayer money:

> We often will hire an at-risk youth who is [sic] 17–18 years old. If you believe that person was heading toward the criminal or social support system, and instead they are now making money and paying taxes, while also saving the government thousands in social supports, that can be an impact of $400,000 per person. (T. Coldwell, personal communication, July 15, 2022)

Chandos is in an interesting position in the supply chain to achieve the SDGs and be impacted in turn by other purchasers who are working toward social and sustainable value outcomes. Chandos can address its own supply chain decisions, increasing the benefits it delivers to communities through its purchasing practices, but the company also benefits from social procurement policies as a B Corp and social purpose business that is able to assist other purchasers to meet their own goals for sustainable development.

Chandos Construction leadership has seen their commitments to social procurement and sustainability rewarded in an increase in contracts. Coldwell shares that in 2021, out of $800 million in total sales, they won approximately $350 million "because of our leadership on this."

He adds that "We were not the lowest fee on approximately $300 million of it," and that "we are winning work we wouldn't have won, with a premium fee, because of the social value we deliver."

One noteworthy example of this phenomenon is the Rundle Affordable Housing Project in Calgary, Alberta, which Director of Business Development Robert Cowan believes Chandos won largely because of the weighted social value criteria in the Request for Proposals (RFP) which was issued in late 2021 (R. Cowan, personal communication, June 14, 2022).

The City of Calgary began implementing Benefit Driven Procurement (BDP) (also referred to as social procurement) in 2019, "seeking to make intentional positive contributions to both the local economy and the overall vibrancy of the community" (City of Calgary, n.d.). On the Rundle Project, the city was asking what proponent general contractors could deliver in terms of diverse hiring, Indigenous inclusion, living wages, and environmental sustainability. "Rundle was the first time I had seen the City of Calgary requesting these things for a major project," Cowan told Buy Social Canada in an interview. "It was encouraging to see a major player like Calgary, that had previously been predominantly decided based on dollars and cents, incorporate social procurement into their procurement practice."

Cowan adds that in recent years, Chandos has seen an increase in social procurement policies and practices in municipalities and government organizations throughout Western Canada and in Ontario, as well as among private clients across Canada. Although they may not all use the language of social procurement, purchasers are increasingly demonstrating that social value outcomes are important to them in both their requests for proposals and interview questions with proponents.

Cowan is hopeful about this trend, reflecting that, "it wasn't that long ago where sustainability wasn't considered, and there was a time when safety wasn't even considered. That was a new question 30 years ago … and now safety is table stakes for construction," and that "social procurement and what companies are doing beyond just building things [...] 30 years from now in my opinion, that will be table stakes."

Policy is a valuable tool to overcome the stigma and myths surrounding social procurement in the construction industry. Cowan has seen cases when companies are forced to practice social procurement as a requirement of their contract on a construction or infrastructure project, and then come to the realization that their preconceptions were wrong. The City of Calgary's BDP Policy, for example, has had a massive impact in the region due to its influential position in the supply chain. Cowan shares that years ago, the social value methods of Chandos Construction were questioned by many. Those who questioned Chandos in the past are taking a second look, as large purchasers such as Calgary award points that require specific outcomes that Chandos is already well positioned to deliver.

"What's encouraging is that people are adopting this, and look at things differently," says Cowan. "Obviously there's a competitive advantage, we're still a business, we're still a construction company, we have 600 employees that need to get paid. But if we can have a competitive advantage that influences the rest of our industry to think about things differently at the same time, and eventually in 10 years if everyone is acting this way, I don't know if there's a greater outcome for it." While Chandos has certainly secured a competitive advantage in a market that increasingly prioritizes social and environmental outcomes, their main intent is not a competitive advantage, Cowan allows: "it's to influence and positively impact our communities through our position in the supply chain." Company President, Tim Coldwell sums it up well:

> Policies create demand and push contractors to practice social procurement. We need the whole industry to focus on it. The big volume buyers of construction making these moves is a great way to push industry to do this and to learn that it's not as hard as they think. (T. Coldwell, personal communication, July 15, 2022)

What often begins with resistance and strictly as a matter of compliance can lead to a change in relationships between market segments. This is currently the case developing in social procurement and community benefit agreements across the construction sector: from fear and myths about the process to engagement, pilots, and early successes. When compliance becomes a market advantage in meeting social value requirements built into RFx and contract deliverables, the industry adapts.

Chandos Construction is not alone in its leadership. Other construction companies are also stepping up to create change through social procurement. Bird Construction, a Buy Social Engage Member, has partnered with Chandos to support the Building Good initiative to innovate the construction sector for increased social impact. Other Buy Social Canada Social Purchasing Partners from the construction sector who have made a commitment to social procurement in their organization and supply chain include Clark Builders, Delnor Construction, and michael + clark construction.

## Looking Ahead

With the integration of social procurement and community benefit agreements in construction, we are transitioning from an industry whose purpose was to merely build structures "on time and on budget" to an industry that is building healthy communities and achieving multiple UN SDGs.

This emerging framework is being led, on the supply side by social enterprise suppliers like EMBERS, and, on the demand side through supportive public policy at all levels of government and by social value purchasers like Chandos Construction. We are witnessing a culture shift across the construction sector. This model uses existing construction contracts and infrastructure investments to achieve the SDGs through employment opportunities, training and apprenticeships, and supply chains that are inclusive of social values: diversity, equity and inclusion.

The challenge, and the imperative, is to continue the journey along this path. To move the story from a few cases like Mike and Neil to a multitude of evidence-based social value goals and measurable results. Each time a social value construction company adds targeted jobs, increases apprenticeships for persons facing barriers to employment, and brings new social enterprises into their supply chain, we have the potential to achieve the UN Sustainable Development Goals, and the opportunity to build healthy communities.

## Bibliography


Atkinson Foundation. (2016). *Making community benefits a reality in Ontario.* https://img1.wsimg.com/blobby/go/155a0c21-72da-4b9c-bddb-4e2a3a316ead/downloads/Atkinson_CBSummary_FA-1-2.pdf?ver=1632423831411

Barraket, J., & Loosemore, M. (2018). Co-creating social value through cross-sector collaboration between social enterprises and the construction industry. *Construction Management and Economics, 36*(7), 394–408.

Bateman, J. (2020). *More British Columbians support community benefits agreements.* Independent Contractors and Businesses Association of BC. https://www.icbaindependent.ca/2020/06/11/news-release-poll-shows-overwhelming-support-for-fair-bidding-in-b-c/

Buy Social Canada. (2019). *Social procurement policy framework.*

Buy Social Canada. (2022). *Voices of industry: A paradigm shift in CBAs.* https://www.buysocialcanada.com/wp-content/uploads/Voices-of-Industry-Paradigm-Shift-in-CBAs.pdf

Canada's Building Trades Unions. (2022). *Community benefits agreements.* Retrieved September 26, 2022, from https://buildingtrades.ca/wp-content/uploads/2022/02/CBA-Report.pdf

Canadian Construction Association. (n.d.). *Workforce.* Retrieved July 25, 2022, from https://www.cca-acc.com/advocacy/critical-issues/construction4cdns/workforce/#:~:text=The%20industry%20is%20facing%20a,Canadian%20Construction%20Association%20(CCA)

Canseco, M. (2020). *More British Columbians support community benefits agreements.* https://researchco.ca/2020/06/01/cba-btc/

Cardus. (2021). *Community benefits agreements: Toward a fair, open, and inclusive framework for Canada.* Retrieved September 26, 2022, from file:///Users/sarahpickering/Downloads/07_14_21_Comm_Benefits_V3.pdf

Cautillo, G. (2020, November 30). *OGCA viewpoint: What are community benefits anyway?* Ontario Construction News. https://ontarioconstructionnews.com/ogca-viewpoint-what-are-community-benefits-anyway-2/

Centre for Social Impact Swinburne. (2021). *Research roundtable on social and indigenous preferential procurement.* https://assets.csi.edu.au/assets/research/Research-Roundtable-on-Social-and-Indigenous-Preferential-Procurement-Summary-Report.pdf

Chandos Construction. (n.d. a) *Net zero 2040.* Retrieved July 25, 2022, from https://www.chandos.com/who-we-are/netzero/

Chandos Construction. (n.d. b) *Our story.* Retrieved July 25, 2022a, from https://www.chandos.com/who-we-are/our-story/

Chandos Construction. (n.d. c) *Social procurement.* Retrieved July 25, 2022b, from https://www.chandos.com/how-we-build/social-procurement/

City of Calgary. (n.d.). *Benefit driven procurement.* Retrieved July 25, 2022, from https://www.calgary.ca/cs/supply/supplyhome/benefit_driven_procurement.html

City of Vancouver. (n.d.). *Community benefit agreements policy.* Retrieved September 23, 2022, from https://vancouver.ca/people-programs/community-benefit-agreements.aspx

Denny Smith, G., Loosemore, M., & Williams, M. (2020). *Assessing the impact of social procurement policies for indigenous people construction management and economics, 38*(12), 1139–1157. https://www.tandfonline.com/doi/full/10.1080/01446193.2020.1795217

Denny-Smith, G., & Loosemore, M. (2017). Integrating indigenous enterprises into the Australian construction industry. *Engineering, Construction and Architectural Management, 24*(5).
Denny-smith, G., Sunindijo, R. Y., Loosemore, M., Williams, M., & Piggott, L. (2021). How construction employment can create social value and assist recovery from covid-19. *Sustainability (switzerland), 13*(2), 1–20. https://doi.org/10.3390/su13020988
Dragicevic, N., & Ditta, S. (2016, April). *Community benefits and social procurement policies: A jurisdictional review*. Mowat Centre. https://www.cardus.ca/research/work-economics/reports/community-benefits-agreements-toward-a-fair-open-and-inclusive-framework-for-canada/
EMBERS Canada. (n.d.). *About us*. Retrieved July 25, 2022, from https://emberscanada.org/about/
EMBERS Staffing. (n.d.). *Embers staffing: A division of Embers*. Retrieved July 25, 2022, from https://embersstaffing.com/about/
GoContractor. *How does construction impact the environment?* (2017, June 21). https://gocontractor.com/blog/how-does-construction-impact-the-environment/
Government of Canada. *Community Employment Benefits Guidance*. (n.d.) Retrieved September 23, 2022, from https://www.infrastructure.gc.ca/pub/other-autre/ceb-ace-eng.html
Governor General's Innovation Awards. (2021). *Embers staffing solutions*. Retrieved July 25, 2022, from https://innovation.gg.ca/winner/embers-staffing-solutions/
Graser, D. (2016). *Community benefits in practice and in policy: Lessons from the United States and the United Kingdom.*
Graser, D. (2018). *Community benefits in York region research report.*
Gunton, C., & Markey, S. (2021). *The role of community benefit agreements in natural resource governance and community development Issues and prospects, 73*. https://doi.org/10.1016/j.resourpol.2021.102152
Gunton, C., Gunton, T., Batson, J., Markey, S., & Dale, D. (2021). *Designing fiscal regimes for impact benefit agreements, 72*. https://doi.org/10.1016/j.resourpol.2021.102004
Inclusive Recovery. (n.d.). *FAQs*. Retrieved July 20, 2022, from https://www.inclusiverecovery.ca/faqs
Kucheran, R. (2020). *Industry perspectives op-ed: Investing in infrastructure must mean investing in Canadians*. https://canada.constructconnect.com/dcn/news/government/2020/06/industry-perspectives-op-ed-investing-in-infrastructure-must-mean-investing-in-canadians
LePage, D. (2014). *Exploring social procurement: Accelerating Social Impact CCC, Ltd.* https://ccednet-rcdec.ca/sites/ccednet-rcdec.ca/files/ccednet/exploring-social-procurement_asi-ccc-report.pdf
Let's Build Canada. (2021, September 9). *New survey finds construction workforce's top election priorities include supports for workers, changes to EI, labour mobility, and transition to Net Zero*. https://www.letsbuildcanada.ca/newsroom/new-survey
Loosemore, M. (2016). Social procurement in UK construction projects. *International Journal of Project Management, 34*(2), 133–144. https://doi.org/10.1016/j.ijproman.2015.10.005
Loosemore, M. (2019). *Social procurement in the construction industry: Challenges and realities.*

Loosemore, M., & Bridgeman, J. (2017a). Corporate volunteering in the construction industry motivations, costs and benefits. *Construction Management and Economics, 35*(10), 641–653. https://doi.org/10.1080/01446193.2017.1315150

Loosemore, M., & Bridgeman, J. (2017b). The social impact of construction industry schools-based corporate volunteering. *Construction Management and Economics, 36*(5), 243–258.

Loosemore, M., & Higgon, D. (2015). *Social enterprise in the construction industry: Building better communities.*

Loosemore, M., & Reid, S. (2019a). The social procurement practices of tier-one construction contractors in Australia. *Construction Management and Economics, 37*(4), 183–200.

Loosemore, M., Alkilani, S., & Mathenge, R. (2020a). The risks of and barriers to social procurement in construction: A supply chain perspective. *Construction Management and Economics, 38*(6), 1–18.

Loosemore, M., Alkilani, S., & Mathenge, R. (2020b). The risks of and barriers to social procurement in construction a supply chain perspective. *Construction Management and Economics,* 1–18.

Loosemore, M., Bridgeman, J., & Keast, R. (2020c). Reintegrating ex-offenders into work through construction a case study of cross-sector collaboration in social procurement. *Building Research & Information, 48*(7). https://doi.org/10.1080/09613218.2019.1699772

Loosemore, M., Denny-Smith, G., Barraket, J., Keast, R., Chamberlain, D., Muir, K., Powell, A., Higgon, D., & Osborne, J. (2020d). Optimising social procurement policy outcomes through cross-sector collaboration in the Australian construction industry Emerald Insight. *Engineering, Construction and Architectural Management, 28*(7).

Loosemore, M., Higgon, D., & Osborne, J. (2020e). Managing new social procurement imperatives in the Australian construction industry. *Engineering, Construction and Architectural Management, 27*(10), 3075–3093.

Loosemore, M., Osborne, J., & Higgon, D. (2020f). *Affective, cognitive, behavioural and situational outcomes of social procurement a case study of social value creation in a major facilities management firm Construction Management and Economics, 39*(3), 227–244.

Loosemore, M., Alkilani, S., & Ahmed, A. (2021a). The job-seeking experiences of migrants and refugees in the Australian construction industry Building Research & Information. *Building Research & Information, 49*(8). https://doi.org/10.1080/09613218.2021.1926215

Loosemore, M., Alkilani, S., & Hammad, A. (2021b). Barriers to employment for refugees seeking work in the Australian construction industry an exploratory study Emerald Insight. *Engineering, Construction and Architectural Management, 29*(2). https://www.emerald.com/insight/content/doi/10.1108/ECAM-08-2020-0664/full/html

Loosemore, M., Alkilani, S., & Murphy, R. (2021c). The institutional drivers of social procurement implementation in Australian construction projects. *International Journal of Project Management, 39*(7), 750–761.

Loosemore, M., Bridgeman, J., Russell, H., & Alkilani, S. Z. (2021d). Preventing youth homelessness through social procurement in construction: A capability empowerment approach. *Sustainability (Switzerland), 13*(6). https://doi.org/10.3390/su13063127

Loosemore, M., Keast, R., Barraket, J., & Denny-Smith, G. (2021f). Champions of social procurement in the Australian construction industry: Evolving roles and motivations. *Buildings, 11*(12), 641. https://doi.org/10.3390/buildings11120641

Macfarlane, R. (2014). *Tackling poverty through public procurement.* www.anthonycollins.com/briefings/social-value-and-public-procurement

Macfarlane, R., & Joseph Rowntree Foundation. (2000). *Using local labour in construction: A good practice resource book.* Policy Press.

MoneyWellWasted.Ca. (n.d.) *Fake CBA – Money well wasted.* Retrieved July 20, 2022, from https://moneywellwasted.ca/fake-cba/

Mustafa, H., & Sengupta, R. (2014). *Connecting capital to sustainable infrastructure opportunities white paper for sustainable infrastructure symposium.*

Office of the Parliamentary Budget Officer. (2022, March 3). *Federal Infrastructure Spending, 2016–2017 to 2026–2027.* https://www.pbo-dpb.ca/en/additional-analyses--analyses-complementaires/BLOG-2122-008--federal-infrastructure-spending-2016-17-2026-27--depenses-federales-infrastructure-2016-2017-2026-2027#:~:text=On%20a%20Public%20Accounts%20(accrual,%2422.8%20billion%20in%202020%2D21

Office of the Procurement Ombudsman. (2020, September). *Social procurement: A study on supplier diversity and workforce development benefits.* https://opo-boa.gc.ca/diversite-diversity-eng.html

Public Services and Procurement Canada. (2020). *Advancing socio-economic goals, increasing competition and fostering innovation—Better Buying—Buying and selling.* https://www.tpsgc-pwgsc.gc.ca/app-acq/ma-bb/posacfi-asgicfi-eng.html

Raiden, A., & Loosemore, M. (n.d.). Social value in the built environment. Routledge & CRC Press.

Raiden, A., Loosemore, M., King, A., & Gorse, C. (2018). *Social value in construction Ani Raiden, Martin Loosemore, Andrew Ki.* 238. https://doi.org/10.1201/9781315100807

Scottish Government. *Community benefits in procurement.* (n.d.) Retrieved September 23, 2022 from https://www.gov.scot/policies/public-sector-procurement/community-benefits-in-procurement/

Toronto Community Benefits Network. (2020). *TCBN stakeholder consultation report for career track in construction project.*

Troje, D. (2021). Policy in practice: Social procurement policies in the Swedish construction sector. *Sustainability (switzerland), 13*(14), 7621. https://doi.org/10.3390/su13147621

Veterans in Construction. (n.d.). Retrieved September 23, 2022, from https://veteransinconstruction.com.au/

Williamson, A. (2021). *Which community benefits agreements really delivered?* Shelterforce.

Yalnizyan, A. (2017). *Community benefits agreements empowering communities to maximize returns on public infrastructure investments.* https://doi.org/10.13140/RG.2.2.34404.42889

# Systemic Approaches to Built Environment Systems

# Relationship Between Changes in Building Culture and the Carbon Footprint from Past to Present: A Case Study from Şanlıurfa-Turkey

Mohammad Ahmad Hussein Khataybeh and Alpay Akgüç

## Introduction

In the wake of a growing population, there is an ever-increasing demand for new building construction, and in turn, for energy resources to meet the needs of occupants. This need increases the consumption of fossil fuels, on which we continue to depend, and this leads to a rise in $CO_2$ emissions. Residential buildings account for approximately 20% of the energy consumption and 9% of $CO_2$ emission in Turkey (International Energy Agency [IEA], 2021).

Turkey meets most of its energy consumption needs from fossil fuels. According to the International Energy Agency (IEA), the energy supply has grown steadily in Turkey to meet the needs of its rapidly growing economy. Despite the recent decline due to the economic crisis in 2018, the total primary energy supply increased by 92% between 2000 and 2019. Fossil fuels dominated the total primary energy supply in Turkey and usage rates have remained stable at approximately 90% since 2000. In 2019, the share of fossil fuels in the total primary energy supply was 83%, the ninth highest rank among the IEA member countries (IEA, 2021).

M. A. H. Khataybeh (✉) · A. Akgüç
Faculty of Architecture and Design, Istanbul Aydin University, Istanbul, Turkey
e-mail: mahmadkhataybeh@stu.aydin.edu.tr

A. Akgüç
e-mail: alpayakguc@aydin.edu.tr

 
T. Walker et al. (eds.), *The Role of Design, Construction, and Real Estate in Advancing the Sustainable Development Goals*, Sustainable Development Goals Series, https://doi.org/10.1007/978-3-031-28739-8_8

Domestic power generation has exploded in recent years, with a growth rate of 59% from 2014 to 2019. This rate was largely associated with power generation from renewable sources, which accounted for 54% of total energy production in 2019. Geothermal power generation, especially, has more than doubled since 2014, accounting for 21% of the total energy production in 2019. Coal production has also increased in recent years following the prior decline during the 2010–2015 period. Coal accounted for 39% of total production in 2019 (IEA, 2021).

$CO_2$ emissions per capita have increased by approximately 46% from 2000 to 2018, despite the economic decline in 2001 and 2013, due to the fact that Turkey meets most of its energy consumption needs from the use of fossil fuels (Fig. 1). $CO_2$ emissions per capita in Turkey were about 2.8 times those of India, 2.4 times those of Brazil, and about 2 times those of Egypt as of 2018 (Fig. 2).

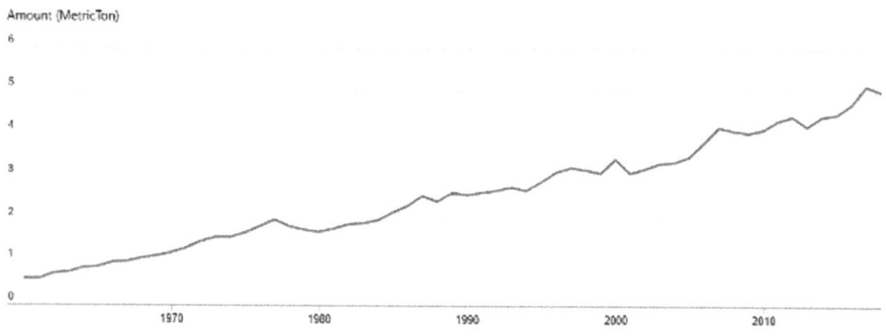


**Fig. 1** $CO_2$ emissions per capita in Turkey (International Bank for Reconstruction and Development [IBRD], n.d.)

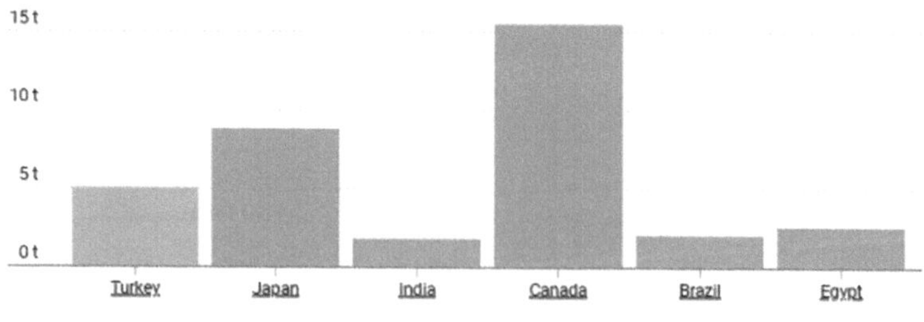


**Fig. 2** $CO_2$ emissions per capita compared to other countries (2018) (IBRD, n.d.)

A review of the change in the carbon footprint of cities across Turkey over the years is key to understanding the difference between Turkey and other countries with regard to $CO_2$ emissions. According to Bozdağ (2021), the carbon footprint dramatically increased between 2014 and 2017 in all the cities across Turkey. The fact that the energy needs associated with the growing number of residential buildings in the Western cities, alongside a rapid boom in urbanization and industrialization, as well as in cities in the east and southeast, were met by fossil fuels, has increased $CO_2$ emissions.

Dependence on fossil fuels also increased in sectors with an increased demand for energy, including manufacturing, transportation, construction, and the dramatic rise in greenhouse gas emissions rendered the climate crisis inevitable. Due to the increased demand for energy, Turkey, as one of the signatories to the Paris Agreement, is committed to use energy, decrease greenhouse gas emissions, and combat climate change more efficiently (Bozdağ, 2021).

The Paris Agreement was ratified in Turkish Law by means of "the Code on the Approval of the Ratification of the Paris Agreement" in 2021. The main objectives of the Agreement are listed below:

- Keep the increase in global average temperature to well below 2 °C above pre-industrial levels and pursue efforts to limit the temperature increase to 1.5 °C.
- Increase the ability to adapt to the adverse impacts of climate change, foster climate resilience, and low greenhouse gas emissions.
- Make finance flows consistent with a pathway toward low greenhouse gas emissions and climate-resilient development (ÖZBEY et al., 2017).

According to Bozdağ (2021), although the $CO_2$ emission of major cities in the Western Turkey reached high levels, they remained almost steady between 2014 and 2017. Nevertheless, the $CO_2$ emissions of Şanlıurfa, a large city located in the southeastern Turkey, which was 25–35 metric tons (Mt) $CO_2$ in 2014, dramatically increased in only a few years to 46–55 Mt $CO_2$ in 2017, i.e., almost double.

In the scope of the present study, passive and active building system energy solutions were investigated with an aim to prevent the dramatic increase in the $CO_2$ emissions in Şanlıurfa, and considering the action plans as stipulated in Goal 7 (accessible and clean energy) and Goal 11 (sustainable cities and communities) among the 2030 Agenda for Sustainable Development goals.

## Change in Carbon Footprint in the Southeastern Turkey

Air pollution has proved to be a serious problem in Turkey due to the industrialization and rapid urbanization that began after the foundation of the republic in Turkey. Şanlıurfa city is one of the locations in Turkey where air pollution has increased significantly in recent years.

The city of Şanlıurfa has become a center for attracting people in the Southeastern Geographical Region with its transition to irrigated agriculture in the 1990s, pursuant to the efforts in the scope of the Southeastern Project (GAP) from the mid-1980s, resulting in rapid population growth. The urbanization rate increased from 50% in 1985 to 92% in 2015. Naturally, such a rapid increase in urbanization in a relatively short time has given rise to a number of environmental and urban problems. Furthermore, industry in the city adversely affects the air quality of the region (Kayan, 2018).

A review by the Turkish Statistical Institute (TUIK) for 2016 reported the electricity consumption per capita in Şanlıurfa was 1482 kWh. Compared to the 2016 data, the same rate increased by 76% in 2020. Therefore, there was a dramatic increase in the carbon footprint of the city of Şanlıurfa (Turkish Statistical Institute [TÜİK], 2020).

Depending on TÜİK data (2020), the electricity consumption per capita in Şanlıurfa increased by approximately 227 kWh each year from 2016 to 2020. The same rate will increase to approximately 3754 kWh per year by 2025 if the trend is maintained. That's more than twice the 2016 level.

## The Effects of Changes in Architectural Culture on Occupants' Comfort

In-depth research was performed with the aim of investigating the comfort levels of residential building occupants vis-a-vis changes in architectural culture in southeastern Turkey. A total of 120 people—comprising 50 new residential building occupants, 50 vernacular house occupants, and 20 high-income house occupants—were surveyed in the city of Diyarbakir. Based on the answers of the respondents, occupant comfort was lower among new residential building occupants compared to the occupants of vernacular residences. Upon reviewing the graphs derived from the responses, it was reported that the indoor performance of the vernacular houses was higher, especially during times when heating was not required (the summer season). While the interiors of the new residential buildings were generally perceived as very hot during the summer months, those of the vernacular houses were perceived as normal. Dissatisfaction with the temperature increased the need for mechanical systems, including air conditioners and air humidifiers, to provide comfortable interior conditions and resulted in an increase in energy expenditures (Dizdar, 2009).

According to Kisa Ovali (2020), the recommended optimum values for bioclimatic comfort conditions for indoors are 19–25 °C, 40–60%, and 0.2–0.4 m/s for indoor air temperature, relative humidity, and max air movement respectively. The further outside of normal values, the greater the detrimental effect on the indoor comfort. This functions alongside increased energy consumption from air conditioning systems in order to provide the occupants with interior conditions to meet desired levels of comfort. As a result, the carbon footprint increases with increased energy consumption where the building culture has adversely changed.

Bioclimatic architecture is based on local architecture, and adopts to the climatic and environmental conditions in the course of creating the interior climate of the building. Climatic effects are kept under control by means of passive architectural components and landscape elements, and thus, optimum indoor comfort is provided in bioclimatic designs. Those climate-responsive designs are not dependent on mechanical systems (active solutions) and such systems feature only in a supporting role (Kisa ovali, 2020).

The present study investigated the climate-responsive building design parameters, in vernacular and new residential buildings in the city of Şanlıurfa. For the purposes of the present study, a rating system was developed to allow scoring for each parameter. The effects of the changes in building culture in Şanlıurfa from past to present on building energy efficiency and building-borne carbon emissions were analyzed based on the above rating system.

## Climate-Responsive Building Design Parameters

There are five different climate zones in Turkey, including hot-humid, hot-dry, moderate-humid, moderate-dry, and cold. The guiding elements for climate-responsive building design parameters are as follows (GEDİK, n.d.):

a. Site selection and adaptation to topography
b. Building orientation
c. Building spacing
d. Building form
e. Building envelope design

In the present study, the building energy efficiency and $CO_2$ emission analyses were carried out on the sample buildings selected from Şanlıurfa, and the climate-responsive building design parameters were assessed based on the hot-dry climatic zone conditions.

## Building Design Parameters of the Hot-Dry Climate Zone

- Locate buildings on the valley floors to avoid excessive solar radiation and protect against the wind that would carry dust.
- Place buildings close to each other and form narrow streets to shade each other in order to protect against the hot wind and sunlight.
- Use water to create ponds or pools and place the same in the courtyard as well as cooling and moistening the environment in the dominant wind direction in the courtyard (Fig. 3).
- Use light colors on the facades.
- Avoid very tall buildings. High-rise buildings are not suitable for hot-dry climatic zones as they are exposed to excessive solar radiation and wind. Instead, design low-rise buildings with closed Western façades or using as few openings as possible.

**Fig. 3** The inner courtyard of Hacibanlar House (Google Maps, n.d.)

**Fig. 4** The building envelope of Hacibanlar House (Google Maps, n.d.)

- Minimize the heat gains and losses by means of building envelope design (thicker walls, materials with high thermal capacity, smaller windows, etc.). Urfa stone, a widely used material especially in vernacular buildings of Şanlıurfa, is a natural type of limestone and was preferred in the construction of buildings in the past as a result of its unique characteristics, including being readily available in nature, able to be processed with vernacular and simple methods, resistant to pressure, and having a high thermal mass (Fig. 4).

  Unique to this region, Urfa stone reflects the identity of the city. As a sustainable building product, Urfa stone is long-lasting, reusable, and recyclable (Öztürk Tel, 2021).

  The thermophysical properties of Urfa stone (limestone) are shown below (Sözen, 2019):
  - Heat conduction coefficient ($\lambda$): 1.261 W/m.K
  - Specific heat ($c_p$): 908 J/kg.°C
  - Density ($\rho$): 2.483 kg/m$^3$
- Reduce the WWR (window to wall ratio) to be protected by excessive solar radiation.
- Design compact buildings with courtyards. The volumes can be shaded, with secondary spaces upon introduction of revaks as in vernacular architecture.

## An Assessment of Case-study Buildings Based on Climate-responsive Building Design Parameters

For the purposes of the study, a rating system was developed to compare the passive system parameters of vernacular buildings to those of the current buildings in Şanlıurfa, a city located in the hot-dry climatic zone in Turkey. Each parameter was scored 1 point on a 10-point scale.

In the scope of the present study, four residential buildings in Şanlıurfa were selected as case-study buildings. Two of the selected buildings were the Hacıbanlar House and Şahap Bakır House, which were vernacular residential buildings well-known in the history of the region. The other two buildings were selected from Şanlıurfa Akabe Multi-Floor Stage I and Şanlıurfa Akabe Low-Rise Vernacular Stage mass housing buildings, which were constructed by the Housing Development Administration (TOKİ) of the Republic of Turkey Ministry of Environment, Urbanization, and Climate Change using contemporary architectural techniques.

### Assessment of Hacıbanlar House

Also known as Ahmet Esmeray House, Hacıbanlar House is located in the Cami Kebir neighborhood in the old city center of Şanlıurfa. The inscription on the keystone of the iwan reads H. 1085 (year 1085 according to Hijri Calendar, and 1674 according to the Gregorian calendar). The Hacıbanlar House was originally built as a residential building with a courtyard, featuring two floors, including a ground floor and a basement, subsequently converted into a culinary museum over the years. In the basement, there is a barn, a cellar, and a haystack (Akkoyunlu, 1989) (Table 1).

**Table 1** Passive design strategies in place in Hacıbanlar House

| No. | Figure | No. | Figure |
|---|---|---|---|
| 1 | 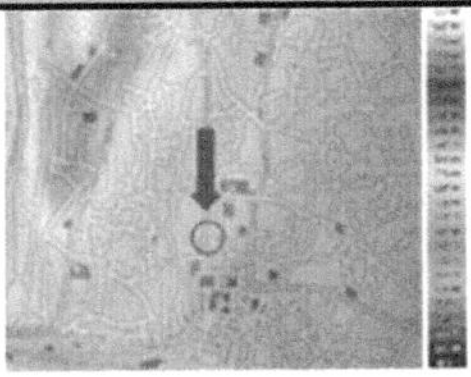 The topography of the surroundings of Hacıbanlar house (Topographic-map, n.d). | 2 |  View of the street in front of Hacıbanları house (Mekan360.com, n.d). |
| 3 |  Satellite image of Hacıbanlar house (Google Maps, n.d.). | 4 |  The courtyard of Hacıbanlar house (Mekan360.com, n.d). |
| 5 |  Basement entrance of Hacıbanlar house (Google Maps, n.d.). | | |

| | |
|---|---|
| • Settlement on the valley floor (see No. 1) | ✓ |
| • Surroundings with narrow street (see No. 2) | ✓ |
| • Compact form (see No. 3) | ✓ |
| • Features courtyard (see No. 5) | ✓ |
| • Wall and floor adjacent to the ground (see No. 5) | ✓ |
| • Light colored façade (see No. 4) | ✓ |
| • Low-rise (see No. 4) | ✓ |
| • The west façade has no window or low WWR (see No. 4) | ✓ |
| • Appropriate materials used in the building envelope (see No. 4) | ✓ |
| • Low WWR (see No. 4) | ✓ |
| Total score of Hacıbanlar House **10/10** | |

## Assessment of Şahap Bakır House

Known as the Arabizade Reşir Efendi House, this house is located in the Pınarbaşı neighborhood in the old city center of Şanlıurfa. There is an inscription in the form of a medallion on the door, which reads H.1192 (year 1192 according to Hijri Calendar, and 1778 according to the Gregorian calendar). Originally built as a residential building, the house was purchased and restored by the Culture, Art, and Publication Board of the Grand National Assembly of Turkey (TBMM). It was named after "TBMM House" and came to be used as a cultural center. The house features two storeys, including the ground floor and the first floor, and features a courtyard. There are two iwans, eight rooms, and a kitchen on the first floor of the house. On the ground floor, there is a barn and three cellars (Akkoyunlu, 1989) (Table 2).

**Table 2** Passive design strategies in place in Şahap Bakır House

| No. | Figure | No. | Figure |
|---|---|---|---|
| 1 | 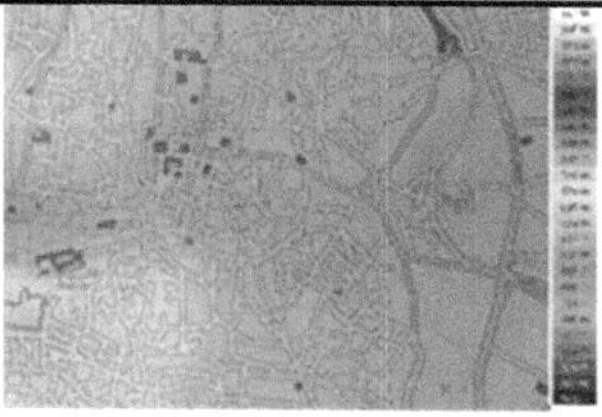 The topography of the surroundings of Şahap Bakır house (Topographic-map, n.d). | 2 |  View of the street in front of Şahap Bakır house (Mekan360.com, n.d). |
| 3 |  Satellite image of Şahapbakır house (Google Maps, n.d.). | 4 |  The courtyard of Hacıbanlar house (Mekan360.com, n.d). |

| | |
|---|---|
| • Settlement on the valley floor (see No. 1) | ✓ |
| • Surroundings with narrow street (see No. 2) | ✓ |
| • Compact form (see No. 3) | ✓ |
| • Features courtyard (see No. 4) | ✓ |
| • Wall and floor adjacent to the ground (see No. 4) | X |
| • Light colored façade (see No. 4) | ✓ |
| • Low-rise (see No. 4) | ✓ |
| • The west façade has no window or low WWR (see No. 4) | ✓ |
| • Appropriate materials used in building envelope (see No. 4) | ✓ |
| • Low WWR (see No. 4) | ✓ |
| Total score of Şahap Bakır House: **9/10** | |

## Assessment of Şanlıurfa Akabe Multi-Floor Stage I Mass Housing Buildings

Located on a sloping site within the Akabe district to the west of Şanlıurfa, these multi-floor apartment buildings were built in 2005. The Şanlıurfa Akabe Multi-Floor Stage I Mass Housing project comprised nine apartment buildings. The buildings with a basement, ground floor, and five more floors above ground level consist of a total of 216 apartments. The total area of each apartment is 128 $m^2$ (Korkmaz, 2006) (Table 3).

**Table 3** Passive design strategies in place in Şanlıurfa Akabe multi-floor stage I mass housing buildings

| No. | Figure | No. | Figure |
|---|---|---|---|
| 1 | 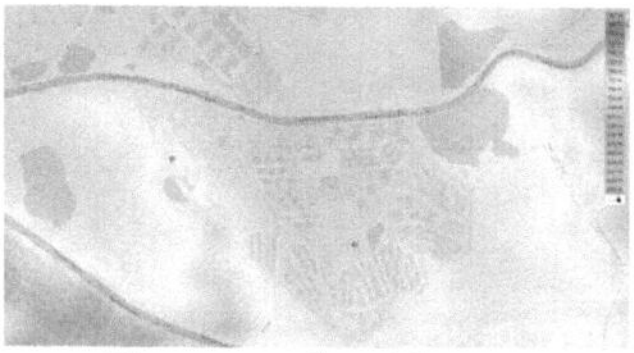 The topography of the area (Topographic-map, n.d). | 2 |  Satellite image of the erea (Google Maps, n.d.). |
| 3 |  view of the west facade of the buildings (Google Maps, n.d.). | 4 |  image showing the building-ground relationship (Google Maps, n.d.). |

| | |
|---|---|
| • Settlement on the valley floor (see No. 1) | ✓ |
| • Surroundings with narrow street (see No. 2) | X |
| • Compact form (see No. 2) | ✓ |
| • Features courtyard (see No. 2) | X |
| • Wall and floor adjacent to the ground (see No. 4) | X |
| • Light colored façade (see No. 1) | ✓ |
| • Low-rise (see No. 1) | X |
| • The West façade has no window or low WWR (see No. 3) | ✓ |
| • Appropriate materials used in building envelope (see No. 3) | X |
| • Low WWR (see No. 3) | X |
| Total Score of Şanlıurfa Akabe Multi-Floor Stage I Mass Housing Buildings: **4/10** | |

## Assessment of Şanlıurfa Akabe Low-Rise Vernacular Stage Mass Housing Buildings

Located to the West of Şanlıurfa Akabe Multi-Floor Stage I Mass Housing project, these traditionally planned low-rise mass housing buildings were erected in 1997. The project is comprised of 790 99-m$^2$ residences with a courtyard. Each residence has two floors and there is a living room, courtyard, and kitchen on the ground floor of the houses, and three bedrooms on the upper floor (Korkmaz, 2006) (Table 4).

**Table 4** Passive design strategies in place in Şanlıurfa Akabe low-rise vernacular stage mass housing buildings

| No. | Figure | No. | Figure |
|---|---|---|---|
| 1 | 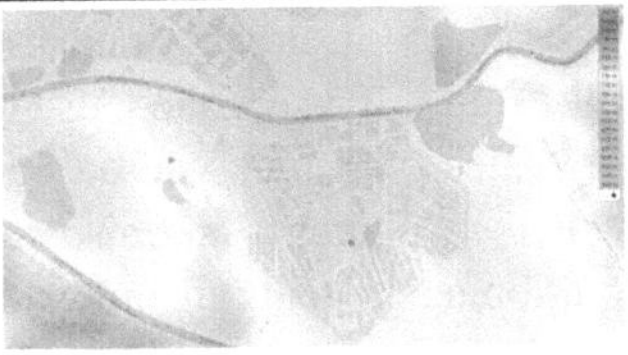 The topography of the area (Topographic-map, n.d). | 2 |  Satellite image of the erea (Google Maps, n.d.). |
| 3 |  Close Satellite image of the erea (Google Maps, n.d.). | 4 |  image showing the building-ground relationship (Google Maps, n.d.). |

| | |
|---|---|
| • Settlement on the valley floor (see No. 1) | X |
| • Surroundings with narrow street (see No. 2) | X |
| • Compact form (see No. 3) | ✓ |
| • Features courtyard (see No. 3) | ✓ |
| • Wall and floor adjacent to the ground (see No. 4) | ✓ |
| • Light colored façade (see No. 3) | ✓ |
| • Low-rise (see No. 1) | ✓ |

| | |
|---|---|
| • The west façade has no window or low WWR (see No. 3) | ✓ |
| • Appropriate materials used in building envelope (see No. 3) | X |
| • Low WWR (see No. 1) | ✓ |
| Total score of Şanlıurfa Akabe Low-Rise Vernacular Stage Mass Housing Buildings: 7/10 | |

## Assessment of Scores

As a result of the score-based assessment of the passive system strategies in place in the selected case-study buildings, the vernacular buildings had higher scores compared to the recently built housing estates.

This result is an indicator of the adverse change in urbanization in terms of efficient use of energy. As a result of this study, it was found that abandoning climate-responsive building design strategies over time caused an increase in the annual heating and cooling loads of the buildings, and also resulted in higher energy consumption from air conditioning systems in order to provide the occupants with bioclimatic comfort. A large increase was observed in the carbon footprint of the city over the years because of the fact that mostly fossil fuels were used to supply energy to the buildings. Furthermore, the results of the present study clearly demonstrate that vernacular buildings were highly successful in reducing environmental impacts by consuming less energy.

## Effect of Electricity Generation with PV Panels on the Annual Fuel Consumption of Buildings

Today, buildings consume approximately 37% of the energy generated and 40% of energy resources across the globe. In addition, buildings account for 40% of the waste in the world. In order to optimize energy use and reduce waste, it is necessary to employ building performance simulation tools and to analyze the role of all building design parameters in terms of comfort, consumption, and cost from the very beginning of the building design (Akgüç, 2017).

The annual electricity consumption per capita in the Şanlıurfa region was 2,618 kW in 2020, according to TUIK data (TÜİK, 2020). Based on the data,

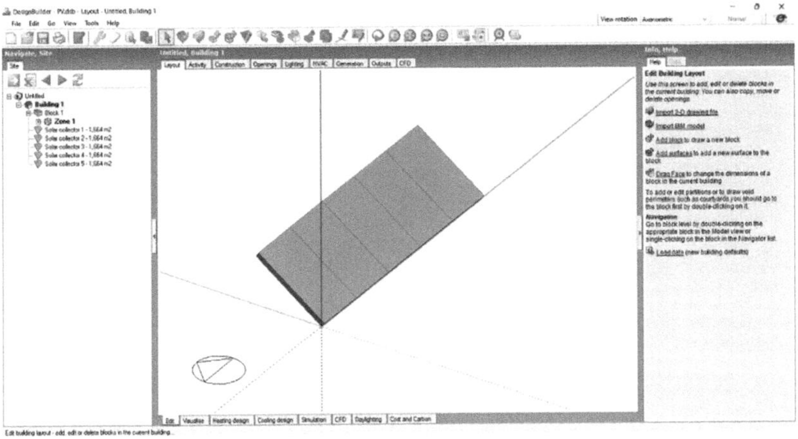

**Fig. 5** Model view of PVs used in the DesignBuilder program

the annual electricity consumption of a Şanlıurfa household can be calculated as 10,472 kWh assuming that a family of four occupies the space.

In the scope of the present study, photovoltaic (PV) panels were modeled using the DesignBuilder building simulation tool, using the detailed dynamic calculation method shown in Fig. 5. The objective was that this simulation model under the Şanlıurfa climatic data, could meet 10% of the annual electricity consumption of a household. The results of the analysis show that the annual amount of electrical energy generated per household would be 1,091.95 kWh, in the case of 5 Uni-Solar PVL-128 brand PVs, inclined at an angle of 35° with respect to the horizontal axis. The technical specifications of the PV panels are shown in Fig. 6.

The simulation concluded that the use of the proposed PV panels was feasible in the Şanlıurfa Akabe Low-Rise Vernacular Stage Mass Housing Buildings, because of the low number of occupants in the building and the size of the roof area was suitable for PV installation. However, PV panel application was not considered a feasible choice for the Şanlıurfa Akabe Multi-Floor Stage I Mass Housing Buildings, due to the high number of occupants and the limited size of roof area per household for electricity generation.

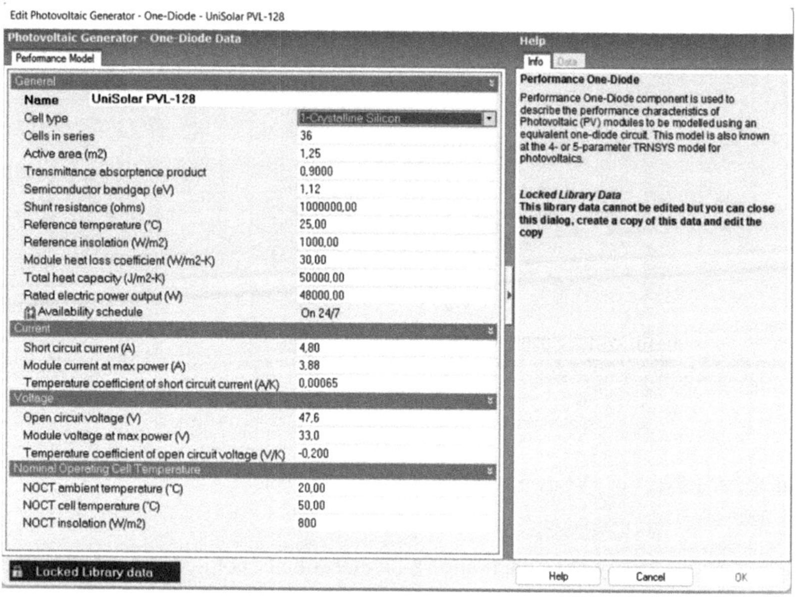


**Fig. 6** Technical specifications of the PVs used in the DesignBuilder program

## Conclusion

Across the globe, buildings are a major source of energy consumption and waste. Every country needs to develop robust policies to ensure building energy efficiency and encourage the use of renewable energy sources, especially given the major adverse environmental impact of fossil fuels. There is a growing need for climate-responsive building strategies that optimize energy consumption.

The present study found that many recently constructed buildings in Şanlıurfa, Turkey, designed without regard to climate, consumed higher amounts of energy in order to provide occupant comfort, compared to vernacular buildings, even though the modern buildings were equipped with more advanced technology compared to vernacular buildings. Furthermore, $CO_2$ emissions increased significantly compared to the past, since most of these buildings used fossil fuels to meet their energy demands.

The rating system as introduced in the present study highlighted the importance of climate-oriented passive system strategies in building design for hot-dry climatic zone and its substantial contribution to the energy efficiency of buildings. In addition, the present study showed that installation of 1.25 $m^2$ PVs on the roof of one of the Akabe Low-Rise Vernacular Stage Mass Housing Buildings could meet 10% of the annual electricity consumption of the building. Higher levels of power generation can be achieved by installing higher capacity PVs in low-rise buildings in Şanlıurfa, provided that further government incentives for renewable energy remain available in Turkey. The increased emphasis on passive system strategies in the Şanlıurfa buildings, and increased incentives for renewable energy in Turkey, would allow for the design of more carbon–neutral buildings in the region and make a major contribution to the implementation by Turkey of the action plans stipulated in Goals 7 and 11 in the 2030 Agenda for Sustainable Development. Reducing greenhouse gas emissions can be achieved through more efficient use of energy resources. However, city planners, architects, engineers, contractors, and subcontractors need to take advantage of collaborative, innovative, and predictive integrated design tools, in order to design buildings that will ensure energy efficiency without compromising the bioclimatic comfort of the occupants.

The primary aim of this research was to contribute to the construction of more energy-efficient and more environmentally friendly buildings based on the rating point system developed as a result of the present study on the basis of climate-responsive building design strategies. The objective was also to introduce a rating system that could form the basis for an algorithm that could serve to reduce the adverse effects of greenhouse gas emissions.

## References

Akgüç, A. (2017). Energy Plus ve Design Builder simülasyon araçlarının işlevi ve kullanımına yönelik genel bir bakış. *TTMD Dergisi, Sayı-11 (Ek).*

Akkoyunlu, Z. (1989). Geleneksel Urfa Evlerinin Mimari Özellikleri (Vol. 13). Kültür Bakanlığı.

Bozdağ, A. (2021). Local-based mapping of carbon footprint variation in Turkey using artificial neural networks. *Arabian Journal of Geosciences, 14*(6), 1–15.

Dizdar, H. (2009). *İklimsel tasarım parametreleri açısından geleneksel ve yeni konutların değerlendirilmesi: Diyarbakır örneği* (Doctoral dissertation, Fen Bilimleri Enstitüsü).

GEDİK, G. Z. (n.d. ). Iklım. https://silo.tips/download/klm-yksek-lisans-prof-dr-glay-zorer-gedk

Google Maps. (n.d.). [Hacibanlar House]. Retrieved October 18, 2022, from https://www.google.com/maps/@37.1507802,38.7899582,3a,90y,203.15h,99.74t/data=!3m7!1e1!3m5!1sAF1QipMmopiXacry6jvUWv7vWWquAmXI0jkBvc_0Zwif!2e10!3e11!7i5376!8i2688

Google Maps. (n.d.). [ŞAHAP BAKIR House]. Retrieved October 18, 2022, from https://www.google.com/maps/@37.1481068,38.7921454,43m/data=!3m1!1e3

Google Maps. (n.d.). [Şanlıurfa Akabe Mass Housing Buildings]. Retrieved October 18, 2022, from https://www.google.com/maps/@37.1382833,38.7392136,1373m/data=!3m1!1e3!5m1!1e4

IBRD—International Bank for Reconstruction and Development. (n.d.). Turkey. https://datacommons.org/place/country/TUR?category=Environment

IEA (2021), Turkey 2021, IEA, Paris https://www.iea.org/reports/turkey-2021, License: CC BY 4.0.

Kayan, A. (2018). Kentleşme Sorunları Kapsamında Şanlıurfa'nın Çevre Sorunları ve Çözüm Önerileri. *Yönetim Bilimleri Dergisi, 16*(32), 299–328.

Kısa Ovalı, P. (2020). Biyoklimatik Tasarım Matrisi (Türkiye). *Trakya Üniversitesi Mühendislik Bilimleri Dergisi.*

Korkmaz, N. M. (2006). *Diyarbakır ve Şanlıurfa'daki toplu konutların kullanım sonrası değerlendirilmesi: Karşılaştırmalı bir analiz* (Master dissertation, Yüksek Lisans Tezi, Dicle Üniversitesi, Fen Bilimleri Enstitüsü).

Mekan360.com. (n.d). [HACIBANLAR House]. Retrieved October 18, 2022, from https://mekan360.com/sanaltur_hacibanlar-evi-mutfak-muzesi-sanliurfa_2107.html

Mekan360.com. (n.d). [ŞAHAP BAKIR House]. Retrieved October 18, 2022b, from https://mekan360.com/sanaltur_meclis-evi-sahap-bakir-evi-8211-isa-beden-evi-s-urfa_2106.html

Özbey, B. G., Geven, F., Güney, K., Bölükbaşi, A., & Günday, B. (2017). Güneydoğu Anadolu Bölgesi'nin hava kalite analizi (Mayıs 2016–2017). *Ankara Üniversitesi Çevrebilimleri Dergisi, 5*(2), 50–64.

Öztürk Tel, H. (2021). Sürdürülebilir Malzeme olan Urfa Taşının Tarihsel Süreçte ve Peyzaj Mimarlığında Kullanımları: Şanlıurfa Örneği. *Journal of Bartin Faculty of Forestry, 23*(3), 742–753.

Sözen, İ. (2019). An approach to the evaluation of vernacular settlements in hot dry climate in terms of thermal comfort: The case of Mardin.

Topographic-map. (n.d). [Turkey topographic map]. Retrieved October 18, 2022, from topographic-map.com

TÜİK—Turkish Statistical Institute. (2020). Electricity consumption per capita (kWh) in Turkey https://elp.tuik.gov.tr/

# Partnership in the Built Environment for Realizing the 2030 Agenda: A Soft Systems Model Incorporating Systems Theory and the Circular Economy

Kamani Sylva and Usha Iyer-Raniga

## Introduction

Human population growth has increased, such that survival needs have surpassed Earth's resource capacity. Demand is increasing and outpacing the biosphere's regenerative and absorptive capacity (Borucke et al., 2013). As an artificial system, a city imports nutrients or resources to sustain its metabolism and generates metabolites or waste (Zhang, 2013). Anthropogenic pressures on the Earth's systems have reached a scale which has destabilized critical biophysical systems, triggering abrupt or irreversible environmental changes that are catastrophic to human well-being (Rockström et al., 2009). Human settlements exploit the planet's virgin resources through linear processes, and the rapid discarding of non-biodegradable matter interrupts natural circular systems at an alarming speed.

Nature offers effective circular systems. The fundamental cycles of water, materials, energy (embodied), and nutrients are classic examples of the circularity of natural subsystems. However, the linear progression of the built environment for human settlements does not mirror the advantages of these circular systems but interferes with their processes (Borucke et al., 2013; Rockström et al., 2009;

K. Sylva (✉)
Department of Engineering Management, Faculty of Engineering, University of Peradeniya, Peradeniya, Sri Lanka
e-mail: kamanisylva@gmail.com; kamanisylva@pdn.ac.lk

U. Iyer-Raniga
RMIT University, Melbourne, Australia
e-mail: usha.iyer-raniga@rmit.edu.au



T. Walker et al. (eds.), *The Role of Design, Construction, and Real Estate in Advancing the Sustainable Development Goals*, Sustainable Development Goals Series, https://doi.org/10.1007/978-3-031-28739-8_9

Zhang, 2013). The industrial revolution in the 1900s gave rise to a notable alteration in the Earth's natural carbon cycle, increasing the amount of carbon in the atmosphere (Nakicenovic, 2004; Riebeek, 2011; Rockström et al., 2009; Sabine et al., 2004). Many factors, such as population increases, economic growth, and energy production, are also projected to contribute to this increase in carbon content (Nakicenovic, 2004).

The consequences of human activities are not limited to alterations in the carbon cycle. Anthropogenic pressures have forced the planet outside the Holocene to the Anthropocene (Rockström et al., 2009). Moreover, modulations of the biology and processes that control carbonate chemistry, such as temperature, alkalinity, or salinity, influence the carbon cycle (Riebeek, 2011). Artificially made surfaces absorb solar radiation and release it at night, allowing temperature differences in cities and their surroundings that nature cannot offset (Nuruzzaman, 2015; Yamamoto, 2006). Furthermore, increasing global average air and ocean temperatures result in the melting of polar ice caps and rising sea levels (Inglezakis et al., 2016). These also affect the hydrological cycle resulting in variations in precipitation and evaporation. The built environment is further responsible for removing vegetation and processing energy, water vapor, and carbon exchange, mainly dominated by a decline in plant growth (Chakravarty & Kumar, 2019). Artificial systems discharge many other non-absorbable substances, making nature's circular processes feeble and leading to catastrophic effects. Attempts have been made at the macro and micro levels to restore the system for a more sustainable outcome. However, these have proved insufficient.

In 1987, the World Commission on Environment and Development, the Brundtland Commission drafted the Brundtland Report, "Our common future", which articulates a commonly accepted definition of sustainable development. Following this, several nations convened at the Rio Summit, also known as the Earth Summit, held in Rio de Janeiro (Brazil) in 1992 to agree upon the Climate Change Convention. This convention led to the Kyoto Protocol and, more recently, the Paris Agreement. The Paris Agreement stipulates a consensus to mitigate global temperature increases to no more than 1.5 degrees centigrade. The Rio summit led to the development of several documents on the Declaration on Environment and Development, Agenda 21, and Forest Principles. A compelling question is whether this documentation has succeeded and proven measurable.

The 178 government delegations of the United Nations at the Rio summit proposed Agenda 21, which covered various aspects of achieving sustainable development through 40 chapters (Chiu, 2012; Langeweg, 1998). They expected to execute the Agenda at local, national, and global levels, with every local government drawing its Agenda 21 to achieve sustainable development by 2000.

However, the plan extended to Rio + 5 (1997), Rio + 10 (2002), Rio + 20 (2012) and finally, the Sustainable Development Summit (2015), which proposed the 2030 Agenda and established the SDGs. There is doubt whether developing countries can achieve the targets of the 2030 Agenda for many reasons, such as poor governance, pandemics, poor usage of available data, and population growth (Adebayo, 2021).

There is an abundance of research publications on sustainable development that discuss solutions for sustainability at the individual and group levels, which help reduce human activity's adverse effects on the biosphere. However, these solutions are fragmented. A systematic solution which considers the entire planet as one system is crucial. Churchman (1968) states that "...problems are interconnected and overlapping. The solution of one has a great deal to do with the solution of another..." (p. 4). According to Churchman (1968), although we have the technological capability to solve these problems, the human world is not organized in a way which promotes the use of these technologies. A systematic change would require partnerships and collaboration between communities, companies, cities, countries, and continents. A partnership that supports human activity without interfering with nature's circular systems would give rise to systems thinking.

This chapter first presents the connectivity of consequences of human activity and their complementary solutions, mimicking the behavior of nature's circular systems. We establish that resolutions to global problems need interconnectedness and that a holistic approach is necessary to handle international issues. We argue that a narrow set of variables to measure isolated phenomena (Seibert, 2018) will only intensify sustainability issues, as the solutions they generate will not address the entire situation. Systems thinking allows the connectivity of subsystems to a whole system with their interwoven relationships. Inputs, throughputs, and outputs of subsystems need connectivity through extensive networks to visualize the effects of one subsystem on another and the system. We incorporate systems theory and circular economy principles throughout the chapter to build a soft systems model for partnerships in the built environment to achieve the 2030 Agenda. According to Robertson (2008), new frontiers for growth are social and psychological rather than technical and economic. We establish that living on a "small and crowded planet" (Robertson, 2008, p. 13) will not be practical without systemic change. Since new growth needs are the foundational tenets of the SDGs, the authors use the language expressed by the SDGs to indicate how the built environment may contribute to the 2030 Agenda through a circular system.

This chapter uses secondary literature to present a soft systems model for the preferred systemic change for the biological community to sustain on the planet.

A soft system subsumes and enhances the difficult-to-follow guidelines in simple interpretations with boxes and arrows (Checkland, 2000). Compared to hard systems, the soft systems approach is more appropriate for ill-defined situations (Checkland, 1990). The proposed model simulates the natural circular scenarios for the performance of the built environment to prevent conflict with ecological systems. Literature showing interconnectivity offers fragmented solutions, which can be connected to provide a visual link to the preferred systemic change to the built environment to realize the 2030 Agenda.

## Systemic Change

There will be many variables involved in enacting systemic future change. Discussions on the foreseeable and preferable futures are not uncommon, but there is still no guarantee of what the future holds. Monitoring and evaluation are the only ways to test what has worked. According to Henchey (1978), professionals who have studied the future use esoteric knowledge or learning from current or past trends to predict the future. Henchey (1978) distinguishes four kinds of futures: what may happen as the possible future, what should happen as the preferable future, what could happen as the plausible future, and what will likely be the probable future. Robertson (2008) presents five views for the future, two of which are most detrimental. The first is where things remain as usual, with no concern for consequences, and the second is when a disaster occurs, whereby catastrophic breakdown may happen. He also suggests authoritarian control as the third, with left and right wings of social stability, and the hyper-expansionist view as the fourth, assuming super industrial drives to break out the present problems could depend on technology. The fifth is the sane, humane, ecological (SHE) perspective, which "appeals to optimistic, participative, reflective people, who reject each of the first four views as unrealistic or unacceptable and believe that a better future is feasible" (Robertson, 2008, p. 13).

According to Robertson (2008), feeling for the future is a mixture of prediction, forecasting with different scenarios, deciding on a preference, planning for the desired future, and action. Building a soft systems model that can realize the 2030 Agenda is necessary for mutual understanding and user-friendly application. A holistic rather than individualistic judgment in a particular niche is preferable in achieving SDG targets. Narrow and specialized assessments could complicate a situation, with expertise inclined to a specific aspect and unable to perceive the consequences of human activity on the whole system (Robertson, 2008). Robertson (2008) is of the view that looking at "piecemeal with tunnel vision" (p. b) could distract attention from the system collapse and limit it to niches such as climate change, global warming, or carbon emissions.

Different stakeholders engage in various methods for predicting future events. Historians study the past to guide the present; physical scientists make predictions based on nature's regularities; economists and sociologists use lived experiences and human affairs (Henchey, 1978). This chapter supports a comprehensive, systemic, and holistic change to realize the 2030 Agenda based on the SHE future. It highlights the indicators essential for sustainability through nature's regularities and supports them through system theories.

## Nature's Circular Systems

The Earth functions as a whole system with the help of natural subsystems or spheres (National Geographic, 2022). The geosphere comprises the lithosphere, hydrosphere, cryosphere, and atmosphere. These spheres interact to provide a sustainable environment for living beings in a narrow zone on the surface of Earth, known as the ecosphere or the biosphere.

There are four fundamental cycles: water, carbon, nitrogen, and oxygen which support life on Earth. These are continuously operating, balancing the planet's spheres to functioning ecosystems. The hydrological cycle, commonly known as the water cycle, plays an overarching role in the functioning of geological and biological Earth, operating beyond the cycling of water to the circulation of solar energy, sediments, and chemical elements for the sustainability of the natural community (Inglezakis et al., 2016; Narasimhan, 2009).

The backbone of life on Earth, carbon, is stored mainly as a rock, with a portion residing in the oceans, atmosphere, plants, soil, and fossil fuels, while living organisms are also made of carbon (Riebeek, 2011). The carbon cycle establishes and nourishes the nutrient cycle through photosynthesis and circulates carbon between reservoirs in the carbon cycle exchange (Riebeek, 2011). Although carbon is sequestered in carbon sinks, the deep ocean, and ocean sediments for long periods, any change in the cycle will result in carbon shifts to other reservoirs (Riebeek, 2011).

Photosynthesis—the process by which organisms use sunlight to synthesize nutrients—originates from the nutrient and energy cycles. This process releases oxygen as a byproduct to provide the energy necessary for life on Earth. Microorganisms decompose the accumulated carbon compounds from the dead bodies of the organisms in the biosphere to release carbon dioxide for circulation.

Nutrient and energy cycles in the biosphere shelter a biological community where the ecosystem operates. The organisms interact with their physical environment within the ecosystem. Nutrient and energy cycles nourish the microorganisms in the biosphere by using natural circular flows of matter. These cycles connect biotic and abiotic components in the ecosystem. Likewise, many natural processes maintain human life's sustainability on Earth. Thus, the biological processes are circular, supporting biotic and abiotic mechanisms on the planet, while reservoirs act as temporary storage facilities for the circulation of stocks.

In a holistic approach, we identify the subsystems' purpose and their interconnectedness to the whole system, circular flow, and the reservoirs as temporary storage facilities as necessary features for the sustainability of the natural system. Theories on systems behavior could support this narrative.

## Systems Thinking

Systems thinking is critical in managing complex problems (Arnold & Wade, 2015). Churchman (1968) identifies systems thinking as the management of subsystems that considers the overall issues and brings up solutions to implement thinking for the whole system. The overarching strategy measures the performance and standards of the subsystems in terms of an overall objective. The entire system's performance depends on the subsystem's management. According to Churchman (1968, p. 8), in systems thinking, a subsystem does not postpone "its thinking until a crisis is reached". In other words, the system and subsystems are always adapting. However, many researchers have tested the concept for its definition. Arnold and Wade (2015) identified three features as critical for systems thinking: the system elements, interconnections, and function (Arnold & Wade, 2015). These three features validate the features identified in natural systems: the elements, and their purpose and interconnectedness, as discussed in the previous section. The system's form has been included with the interconnections and the feedback, whereas the dynamic behavior reflects the stocks, flows, and variables, and identifies non-linear relationships (Arnold & Wade, 2015). The stocks, flow, and variables recognize the temporary storage, sinks, or reservoirs in natural systems (Riebeek, 2011).

A city in the built environment could be a model to represent a subsystem or an element to reach the goals of the 2030 Agenda, as depicted in Fig. 1. However, a city does not function in isolation. Coexisting systems collide and function to produce solutions to the whole system, the built environment, and

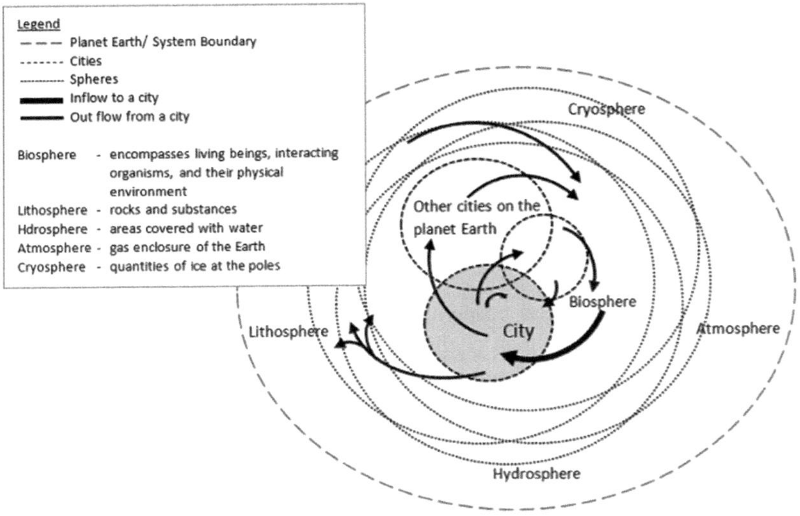


**Fig. 1** Proposed circular thinking for a city (*Source* Authors)

the planet. Problems extending beyond geographical limits with their unknown political interference and geopolitical boundaries will necessitate regional and global approaches for the sustainability of the whole system (Langeweg, 1998). A city must focus on its input, throughput, and output for co-existence. The linking of subsystems is vital for achieving the 2030 Agenda for the entire system to survive without compromising the natural systems. This linking presents the interconnectedness of elements and the system's interference with a city, as illustrated in Fig. 1. A city is an open system linking the Earth's subsystems for its inputs and outputs. This process must take a circular form whereby the speed of the production process should match nature's circular flow of material.

"Circular Thinking" with a whole system approach is required for a city to remain sustainable while maintaining an ecological balance, as opposed to the linear thinking that prevails at present. Predominantly, the linear thinking currently seen in cities exploits resources and produces waste without identifying the consequences of these actions on the overall system (i.e., on the biosphere). However, it is necessary to identify and relate its connections to naturally occurring processes. The built environment processes should align with the natural circular processes. For example, vernacular settlements in Sri Lanka practiced co-existence with nature in traditional farming. Forest reservations for water conservation and providing allowances for wildlife habitats are dominant features in village planning in these settlements. At the same time, activities such as building shelters and transportation have been managed with highly eco-friendly materials and practices (Sylva & Sylva, 2021) with a higher-order circularity (Potting et al., 2017).

## The Circular Economy

The concept of the circular economy intends to correspond to natural circular systems and merge rather than conflict with them. The idea helps to reduce the extraction of virgin material and reuse or recycle the waste of linear processes. This can be achieved by tying up the ends of the linear system to create a circular process. The shift from a linear system impacts the design itself and the flow of material, energy, water, and other materials used throughout the product's life. The 10R Framework proposed by Cramer (2017) suggests that circular systems could prevent waste, and potential value may be retained or enhanced by following the R principles for products: Refuse, Reduce, Renew, Reuse, Repair, Refurbish, Remanufacture, Repurpose, Recycle, and Recover. Among the several circularity strategies, the 10R Framework prefers more intelligent product manufacturing and product sharing as higher-order approaches than lower-order plans (Cramer, 2017; Potting et al., 2017). The lower circularity approaches are recycling and incineration to recover energy (Potting et al., 2017). Environmental benefits will be proportional to the order of the circularity.

The pragmatism of circular metabolism depends on how the system's elements connect to function (Arnold & Wade, 2015). Isolating a subsystem or any individual of a system affects the circular flow. The circular design will be effective once each subsystem is a member of the circular supply chain of the whole system. However, it is necessary to identify a strong sustainability concept which assumes clear limits or stocks of the ecological systems which may not be exceeded for the tradeoff of economic growth or social well-being (Langeweg, 1998; Sylva, 2020a). Partnerships will ensure sharing of resources among the stakeholders. We propose a model of collaborations to reflect elements and their interconnections with their purpose.

A circular economy's foundation is built on ecology, systems thinking, and environmental economics, describing how the resource supply chain can be closed on a local, national, or global scale. Furthermore, several federal governments have promoted the circular economic concept as an approach to monetary gain

(Korhonen et al., 2018). However, a distinct methodology to incorporate systems thinking and circular economy globally have yet to be formulated. This chapter focuses on the linkage through partnerships in the built environment using a soft systems model.

## Soft Systems Model for Partnerships Incorporating Systems Theory and the Circular Economy

To realize the overarching goals of the total system, systems thinking requires the execution of actions at the subsystem level with an understanding of overlapping needs. In applying developmental approaches to a subsystem, such as cities, countries, and continents, physical geographical control may not always be the limit for thinking, planning, and management. Gerten et al. (2013) identify boundaries scaling up to the global domain in the water cycle. They propose a limit for fresh water and combine bottom-up and top-down approaches to maintain the hydrological cycle with its ecosystem's sustainability. In addition, they also consider cascading impacts from rivers to the coast in decision-making within the physical boundaries of a city. Literature has highlighted prospering attempts in ancient water management methods whereby the thinking has gone beyond a village's physical limitations (Sylva & Sylva, 2021). The cascading water management system connecting village systems to the whole system is a classic example in Sri Lanka. Ensuring the discharge necessary to prevent pollution of the subsequent systems can be found in the introduction of natural water cleaning ponds at the exit of one subsystem. This depicts connectivity and whole-system thinking. In Sri Lanka, people in vernacular settlements built large reservoirs in the dry zone interconnected with cascading small village tanks to maintain water supply throughout the dry seasons for agriculture (Sylva & Sylva, 2021).

The sustainability of the city as a system could adopt two ideological paradigms for the structure, top-down and bottom-up, whereby expert-led or community-based decisions are applied in decision-making, respectively (Khadka & Vacik, 2012; Kwok et al., 2016; Moon, 2017; Reed et al., 2006). Debates on the top-down and bottom-up approaches to sustainability identify that policymaking and leadership are far more effective in top-down approaches. "Individuals can be quite agile, given the license to be. But without overarching goals, agility is random and unsustainable" (Moon, 2017). Top-down measures are more quantitative and easily comparable, and national or international scales on a macro level are used (Reed et al., 2006). However, top-down approaches seek quick-fix solutions to complex problems without contextual verification at the grassroots level

(Khadka & Vacik, 2012; Reed et al., 2006). Instead, the method of application or implementation has been effective in bottom-up approaches. The community could develop or accommodate gauges to measure their standards to fit the overarching objectives (Kwok et al., 2016). Adult education learning and social activism are apparent in bottom-up approaches, with a maintained understanding of the local context (Reed et al., 2006). A city as a subsystem needs to focus on SDGs as top-down policies and build context-specific bottom-up approaches to realize the 2030 Agenda. The connectivity should then be easily identifiable for implementation.

The 2030 Agenda expects a global partnership for solidarity, with emphasis on the needs of the poorest and most vulnerable through "the participation of all countries, all stakeholders and all people" (United Nations [UN], 2015, p. 2). Everyone should be a member of at least one subsystem connected to the whole system (i.e., the biosphere). We depict the expected or preferred connectivity of the complex relationships in Fig. 2 through a soft systems model. This model follows the approach recommended by Cramer (2017) to "think global, act local" through the 10R Framework. Providing statistical data on SDG indicators is weak and limited to high-income countries (Kitzmueller et al., 2021). A soft systems model will ensure the distinct relationship needed for realizing the 2030 Agenda while allowing opportunities to systematically report the progress of identified indicators through responsible bodies in the entire system.

The 2030 Agenda and the SDGs should act as the Global Policy Guideline and Direction in the circular supply chain as shown in Fig. 2. In applying systems theory, the system's function is thus clear to all actors or members of the system. The 10R framework is proposed to be used top-down in the decision-making process for the products of the built environment, with a greater focus on more innovative products for society (Potting et al., 2017). Multi-National Corporations (MNCs) and Small and Medium Enterprises (SMEs) act as the circular supply chain partners for built environment materials in the soft system model presented in Fig. 2. The global production processes through MNCs should adhere to the 10R Framework of circular economy, considering the higher-order circularity concepts as their focus but not neglecting the allowance for the lower-order concepts (Potting et al., 2017). MNCs should first adopt concepts such as refusing and rethinking to avoid products that may cause environmental damage and offer alternatives, according to the SHE paradigm. They should also reduce virgin material extraction by improving processes and products efficiency. The MNCs should present a set of indicators to accommodate reusable material to align in the reverse supply chain to allow for lower-order circularity concepts such as recycling the waste

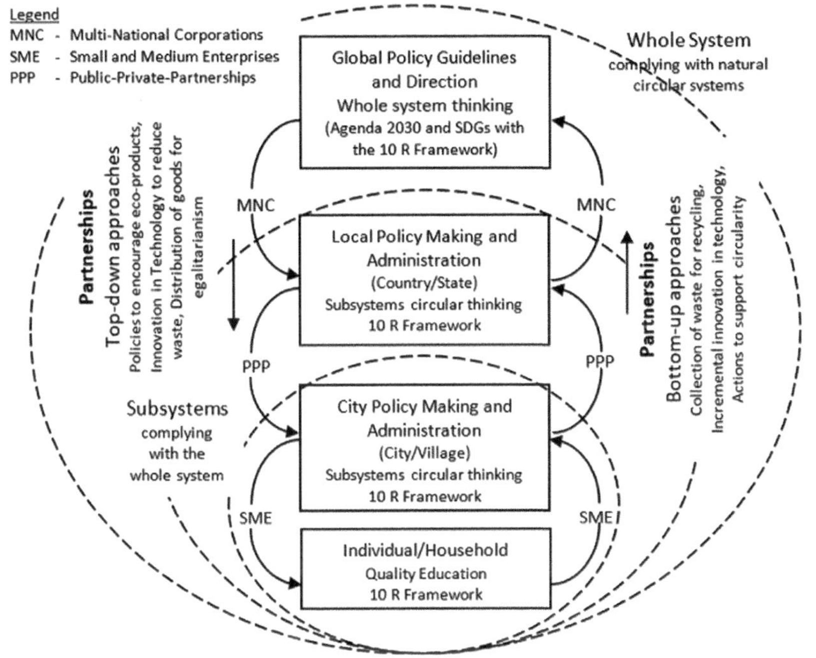


**Fig. 2** Soft systems model for a global circular supply chain for the built environment (*Source* Authors)

products necessary for systemic change. Reusing, repairing, refurbishing, and repurposing are intermediatory-level circularity concepts (Potting et al., 2017).

SMEs should cater to the intermediatory level by introducing relevant industries, such as repair shops while becoming the MNCs' sales agents and raw material suppliers. This partnership would ensure the sustainability of SMEs in the market rather than diminishing with market forces created by the MNCs. MNCs should be made dependent on SMEs to collect waste to resources for their supply chain through relevant policies designed under a Corporate Social Responsibility (CSR) guideline. The governing bodies should create the interconnection of these players or the elements of the system through Public–Private-Partnerships (PPPs) while maintaining the appropriate indicators to measure progress. Figure 3 illustrates the proposed arrangement of partnerships.

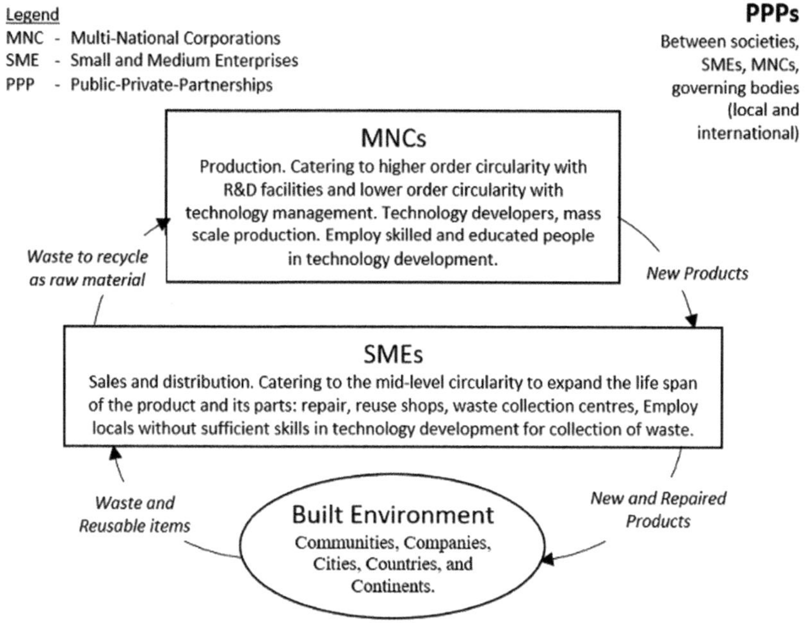


**Fig. 3** Proposed arrangement of PPPs (*Source* Authors)

A radical technological change is not what is expected at this level for the transition to a circular economy in the built environment; a strong PPP which helps the survival of both MNCs and SMEs, is required. If the city administration cannot recycle materials, it is expected to build strong PPPs to attend to this need with local or international links. Moreover, the terms and conditions of the local administration with MNCs who provide materials for the built environment are expected to include agreements to collect all possible materials for recycling under strong CSR policies and guidelines (Somachandra et al., 2021). The circulation of materials within the cities is expected to be through SMEs. This partnership ensures the sustainability of SMEs as employment providers to societies in the SHE model.

Circular economic transitions do not always need a fundamental technological change (Potting et al., 2017). "High-level circularity strategies more often require socio-institutional changes throughout the product chain and innovation in product design and revenue model, whereas low-level strategies rely more on technological innovation" (Potting et al., 2017, p. 7). Socio-institutional changes may need rules and regulations intact and inclined toward building a fair circulation process.

Rules, regulations, host government interventions, and related infringements have pushed MNCs toward developing countries where no or minimal strict policies are set to prevent detrimental production (Doz & Prahalad, 1980; Sylva, 2020b). Most MNCs were misusing the willingness of the developing nations to acquire their investments by exploiting both skilled and non-skilled workforces (Ferdausy & Sahidur, 2009; Sylva, 2020b). A solid policy incorporating a CSR component is necessary to establish good partnerships with SMEs and prevent such exploitation (Sylva, 2020b). The public sector should encourage non-state partners to build a circular supply chain and contribute to the common benefit through profit-sharing (Sylva, 2020d). The private sector should "support the system through fair taxation policies for the usage of public infrastructure, provide employment opportunities to the community, and share profit among stakeholders for equity while adhering to environmental regulations as a part of their responsibility" (Sylva, 2020d, p. 10) for the anticipated systemic change.

The thinking process requires rectification in order to successfully incorporate systemic change. Product developers play a significant role in deciding on alternative products. However, the responsibility cannot be limited to the developer. The consumer should be ready to accept alternative products with higher circularity. The change in consumer behavior in the built environment is to be achieved through quality education which internalizes values of sustainability (Sylva, 2020c). Waste reduction, reuse, and recycling are to be taught by incorporating emerging concepts such as deconstruction, recyclability, and Design for Disassembly (DfD) (Rios et al., 2015; Soh et al., 2015) at the required levels. The construction of the built environment must focus on the 'construction in reverse' or deconstruction to minimize waste dumps using new terminology for a new practice (Rios et al., 2015). Through deconstruction, it is expected to reuse all possible parts or products for new constructions, reducing landfills and virgin material at the level of demolition. Design for Disassembly encourages an easy-to-dissemble mindset and an optimal disassembly sequence based on practical considerations (Soh et al., 2015). Recycling will involve both upcycling and downcycling, minimizing the dumping of waste as much as possible to close the loop of the material cycle in the built environment (Kibert & Chini, 2000; Soh et al., 2015). Aligning the distributor, the vendor, the buyer, and the user in the circular process will lead to realizing the SDGs.

## Potential Pitfalls

The effects of the soft systems model on the SDGs and the realization of the 2030 Agenda could be considered for any city in the world. Closing the ends of linear systems requires both local and international connectivity. The inculcation of institutionalized norms through policies (Sylva, 2020b) and internalized values through quality education (Sylva, 2020c) are essential requirements of the expected change. Education through practical exposure to sustainable development will develop the necessary individuals' psychomotor skills to cater for circularity (Sylva, 2020c). Rejecting harmful materials and materials that are not recyclable (Cramer, 2017; Potting et al., 2017) while encouraging the use of recycled materials (lower order circularity) (Cramer, 2017; Potting et al., 2017) will lead to responsible consumption and production. However, if systems are corrupt, weak institutions with weak paradigms (Sylva, 2020d), the model's effectiveness is in question.

All other SDGs can be identified as indicators for development at different levels of the circular supply chain. MNCs and SMEs will follow the guidelines at the registration level as business entities. They will be bound to provide statistical and qualitative data on achieving the target through their partnerships. Assigning responsibilities to identified members in the supply chain will reduce the difficulties identified in data gathering (Kitzmueller et al., 2021) and other problems that developing countries may face (Adebayo, 2021) in implementing the 2030 Agenda. Each player in the circular design will focus on the system to meet goals such as clean water, sanitation, and affordable clean energy. Decent work and economic growth should be considered within the throughput of the built environment while supplying adequate employment opportunities for people. The skilled workers employed by MNCs to create alternatives to the harmful and non-skilled people used effectively by SMEs to collect waste for recycling could provide significant employment opportunities. Both sectors should get equal attention so that fair treatment and equitable remuneration will be given to those engaged. However, there may be divisions among employees for the required skill and qualifications, which quality educational practices should eliminate.

The built environment should focus on regenerative practices such as allowance for parks and natural water cleaning ponds and preserving connectivity of the ecosystem resources as their CSR requirements. City planning should provide flexibility for regeneration, considering the higher-level circularity strategies. The output of the built environment could then enrich the priorities for terrestrial and marine life on the planet through regenerative ecosystems. Good health and well-being could be ensured while supporting wildlife for an ecological balance to combat climate actions. Providing a decent built environment and opportunities for employment for all through production, distribution, sales, construction, operation and maintenance, demolition, and circulation, will reduce inequalities and poverty with equal opportunity for all, thus leading to zero hunger. Nevertheless, affluent societies will prefer private space in comparison to the upliftment of public space. The government's intervention will be needed in some contexts to ensure its citizens' equity.

Incorporating men and women in all activities will lead to a gender balance in the work environment and a balance in power relationships in a non-work environment. Collaborating with institutions is key to representing stakeholder relationships at the appropriate stage of the built environment's life cycle, leading to less confusion and better alignment among the institutions to encourage the adoption of peace, justice, and strong institutions. The concepts of the circular economy should organize the built environmental partnerships with relevant indicators to add value to ecosystem resources to realize the 2030 Agenda with a holistic view. Regular and systematic monitoring of indicators will be essential, which may need additional costs for operations in some cities. The monitoring process will also open new employment opportunities for the citizens.

## Conclusion

This chapter has focused on developing the conceptual framework for a soft systems model for the built environment to realize the expected outcomes of the 2030 Agenda. The model is based on understanding the importance of the Earth as a system, the role of systemic change, and recognizing the inherent nature of circular systems that occur in the natural world. The principles of a circular economy underlie the systems thinking approach to create the model and rely on partnerships for implementation in a systematic manner.

The application of the model may require strong policies and the commitment of institutions to the circularity of materials. Weak and corrupt systems may have difficulties in aligning the necessary partnerships for the successful application of the model. However, quality education could overcome these barriers. It is essential to investigate further how cities could adopt and adapt the proposed model for the realization of the 2030 Agenda by focusing on a pilot project. Strong CSR policies for MNCs may lead to the initiation of the process throughout the world to combine the whole world into one circular supply chain to enable reaching targets set by the SDGs for 2030 and beyond.

## References


Adebayo, A. (2021, June 8). *Developing Countries and Sustainable Development Goals (SDGs): Can they achieve the 2030 Agenda?* LinkedIn. Retrieved June 20, 2022, from https://www.linkedin.com/pulse/developing-countries-sustainable-development-goals-sdgs-ambrose/

Arnold, R. D., & Wade, J. P. (2015). A definition of systems thinking: A systems approach. *Procedia Computer Science, 44*, 669–678. https://doi.org/10.1016/j.procs.2015.03.050

Borucke, M., Moore, D., Cranston, G., Gracey, K., Iha, K., Larson, J., Lazarus, E., Morales, J. C., Wackernagel, M., & Galli, A. (2013). Accounting for demand and supply of the biosphere. *Ecological Indicators, 24*, 518–533. https://doi.org/10.1016/j.ecolind.2012.08.005

Chakravarty, P., & Kumar, M. (2019). Chapter 6 - Floral species in pollution remediation and augmentation of micrometeorological conditions and microclimate: An integrated approach. In V. C. Pandey & K. Bauddh (Eds.), *Phytomanagement of polluted sites* (pp. 203–219). Elsevier. ISBN 9780128139127. https://doi.org/10.1016/B978-0-12-813912-7.00006-5

Checkland, P., & Scholes, J. (1990). *Soft systems methodology in action.* Wiley.

Checkland, P. (2000). *Systems research and behavioral science* (Vol. 17). John Wiley & Sons, Ltd.

Chiu, R. (2012). Sustainability. *International Encyclopedia of Housing and Home, 91–96.* https://doi.org/10.1016/b978-0-08-047163-1.00688-3

Churchman, C. W. (1968). *The systems approach.* Dell Publishing Company.

Cramer, J. (2017). The raw materials transition in the Amsterdam Metropolitan Area: Added value for the economy, well-being, and the environment. *Environment: Science and Policy for Sustainable Development, 59*(3), 14–21. https://doi.org/10.1080/00139157.2017.1301167

Doz, Y., & Prahalad, C. K. (1980). How MNCs Cope with host government intervention. Retrieved January 18, 2019, from https://hbr.org/1980/03/how-mncs-cope-with-host-government-intervention

Ferdausy, S., & Sahidur, R. (2009). Impact of multinational corporations on developing countries. Chittagong University Journal of Business Administration (Vol. 24) pp. 111 137. ISSN: 2219–4843.

Gerten, D., Hoff, H., Rockström, J., Jägermeyr, J., Kummu, M., & Pastor, A. V. (2013). Towards a revised planetary boundary for consumptive freshwater use: Role of environmental flow requirements. *Current Opinion in Environmental Sustainability, 5*(6), 551–558. https://doi.org/10.1016/j.cosust.2013.11.001

Henchey, N. (1978). Making sense of future studies. *Alternatives, 7*(2), 24–27. http://www.jstor.org/stable/45030200

Inglezakis, V., Poulopoulos, S., Arkhangelsky, E., Zorpas, A., & Menegaki, A. (2016). Aquatic environment. *Environment and Development, 137–212.* https://doi.org/10.1016/b978-0-444-62733-9.00003-4

Khadka, C., & Vacik, H. (2012). Comparing a top-down and bottom-up approach in the identification of criteria and indicators for sustainable community forest management in Nepal. *Forestry, 85*(1), 145–158. https://doi.org/10.1093/forestry/cpr068

Kibert, C. J., & Chini, A. R. (2000). Overview of deconstruction in selected countries, University of Florida. ISBN 0-9643886-3-4.

Kitzmueller, I., Stacy, B., & Mahler, D. G. (2021, August 10). Are we there yet? Many countries don't report progress on all SDGs according to the World Bank's new Statistical Performance Indicators. Retrieved June 22, 2022, from https://blogs.worldbank.org/opendata/are-we-there-yet-many-countries-dont-report-progress-all-sdgs-according-world-banks-new

Korhonen, J., Honkasalo, A., & Seppälä, J. (2018). Circular economy: The concept and its limitations. *Ecological Economics, 143*, 37–46. https://doi.org/10.1016/j.ecolecon.2017.06.041

Kwok, A. H., Doyle, E. E., Becker, J., Johnston, D., & Paton, D. (2016). What is 'social resilience'? Perspectives of disaster researchers, emergency management practitioners, and policymakers in New Zealand. *International Journal of Disaster Risk Reduction, 19*, 197–211. https://doi.org/10.1016/j.ijdrr.2016.08.013

Langeweg, F. (1998). The implementation of Agenda 21 our common failure. *Science of the Total Environment, 218*(2), 227–238. https://doi.org/10.1016/s0048-9697(98)00210-1

Moon, F. (2017, June 27). Sustainability: Top down or bottom up? - Expressworks International. Retrieved May 10, 2022, from https://www.expressworks.com/health-safety-environmental-programs/sustainability-top-down-or-bottom-up/

Nakicenovic, N. (2004). Socioeconomic driving forces of emissions scenarios. In *The global carbon cycle* (pp. 225–239). Island Press.

Narasimhan, T. (2009). *Encyclopedia of Inland Waters*. Academic Press.

National Geographic. (2022). Earth's Systems. Earth's Systems | National Geographic Society

Nuruzzaman, M. (2015). Urban Heat Island: Causes, effects and mitigation measures—A review. *International Journal of Environmental Monitoring and Analysis, 3*(2), 67–73. https://doi.org/10.11648/j.ijema.20150302.15

Potting, J., Hekkert, M., Worrell, E., & Hanemaaijer, A. (2017). *Circular economy*. PBL Netherlands Environmental Assessment Agency.

Reed, M. S., Fraser, E. D., & Dougill, A. J. (2006). An adaptive learning process for developing and applying sustainability indicators with local communities. *Ecological Economics, 59*(4), 406–418. https://doi.org/10.1016/j.ecolecon.2005.11.008

Riebeek, H. (2011, June 16). The Carbon Cycle—NASA. Retrieved May 10, 2021, from https://earthobservatory.nasa.gov/features/CarbonCycle/page1.php

Rios, F. C., Chong, W. K., & Grau, D. (2015). Design for Disassembly and deconstruction—Challenges and opportunities. *Procedia Engineering, 118*, 1296–1304. https://doi.org/10.1016/j.proeng.2015.08.485

Robertson, J. (2008). *The Sane Alternative*. London, James Robertson. Retrieved June 22, 2022, from www.jamesrobertson.com/book/thesanealternative.pdf

Rockström, J., Steffen, W., Noone, K., Persson, Å., Chapin III, F.S., Lambin, E., Lenton, T.M., Scheffer, M., Folke, C., Schellnhuber, H., Nykvist, B., De Wit, C.A., Hughes, T., van der Leeuw, S., Rodhe, H., Sörlin, S., Snyder, P.K., Costanza, R., Svedin, U., Falkenmark, M., Karlberg, L., Corell, R.W., Fabry, V.J., Hansen, J., Walker, B., Liverman, D., Richardson, K., Crutzen, P. & Foley, J. (2009). Planetary boundaries: Exploring the safe operating space for humanity. Ecology & Society, *14*(2), 32. http://www.ecologyandsociety.org/vol14/iss2/art32/ [online].

Sabine, C. L., Heimann, M., Dorothee, P. A., Bakker, C. E., Chen, C. A., Christopher, B., Gruber, F. N., Quere, C. L., Prinn, R. G., Richey, J. E., Lankao, P. R., Sathaye, J. A., & Valentini, R. (2004). Current status and past trends of the global carbon cycle. In C. B. Field & M. R. Raupach (Eds.), In *The global carbon cycle: integrating humans, climate, and the natural world* (pp. 17–43). SCOPE Series. Island Press. ISBN: 9781559635271, 1559635274.

Seibert, M. (2018, November 8). Systems thinking and how it can help build a sustainable world: A beginning conversation. Retrieved May 15, 2022, from https://mahb.stanford.edu/blog/systems-thinking-can-help-build-sustainable-world-beginning-conversation/

Soh, S., Ong, S., & Nee, A. (2015). Application of design for disassembly from remanufacturing perspective. *Procedia CIRP, 26*, 577–582. https://doi.org/10.1016/j.procir.2014.07.028

Somachandra, W. D. I. V., Sylva, K. K. K., & Dissanayake, P. B. R. (2021). Strengthening sustainability in Sri Lankan construction industry: Through corporate social responsibility. In U. Iyer-Raniga (Ed.), In *Sustainability in the built environment in the 21st century: Lessons learned from India and the region. SDRAP 2019*. Environmental Science and Engineering. Springer. https://doi.org/10.1007/978-3-030-61891-9_14

Sylva, K. K. K. (2020a). A value for the non-valued: Valuation of ecosystem resources. In S. Patti, & G. Trizzino (Eds.), *Advanced integrated approaches to environmental economics and policy: Emerging research and opportunities* (pp. 49–70). IGI Global. https://doi.org/10.4018/978-1-5225-9562-5.ch003

Sylva, K. (2020b). Global policymaking process of inequality reduction: Equalities and non-equalities of inequality. In W. Leal Filho, A. Azul, L. Brandli, P. Özuyar, & T. Wall (Eds.), *Reduced inequalities. Encyclopedia of the UN Sustainable Development Goals*. Springer. https://doi.org/10.1007/978-3-319-71060-0_50-1

Sylva, K. (2020c). Proficiency for assessment in quality education: Internalization of values of sustainability. In W. Leal Filho, A. M. Azul, L. Brandli, P. G. Özuyar, & T. Wall (Eds.), *Quality education. Encyclopedia of the UN Sustainable Development Goals*. Springer. https://doi.org/10.1007/978-3-319-95870-5_73

Sylva, K. (2020d). Regulatory paradigm shift for social peace and justice. In W. Leal Filho, A. M. Azul, L. Brandli, A. Lange Salvia, P. G. Özuyar, & T. Wall (Eds.), *Peace, justice and strong institutions. Encyclopedia of the UN Sustainable Development Goals*. Springer. https://doi.org/10.1007/978-3-319-71066-2_59-1

Sylva, K. K. K., & Sylva, K. K. L. A. (2021). Sustainable city planning and management strategies in vernacular settlement patterns in Sri Lanka. In U. Iyer-Raniga (Ed.), *Sustainability in the built environment in the 21st century: Lessons learned from India and the region. SDRAP 2019*. Environmental Science and Engineering. Springer. https://doi.org/10.1007/978-3-030-61891-9_9

United Nations [UN]. (2015). Resolution adopted by the General Assembly on September 25 2015, A/RES/70/1.

Yamamoto, Y. (2006). Measures to mitigate urban heat islands. *Science and Technology Trends Quarterly Review, 18*(1), 65–83.

Zhang, Y. (2013). Urban metabolism: A review of research methodologies. *Environmental Pollution, 178*, 463–473. https://doi.org/10.1016/j.envpol.2013.03.052

# A Sustainable Approach to the Planning, Organization, and Management of Big Events in the Music Entertainment Industry

Marco Mancini

## Introduction

The occurrence of the COVID-19 Pandemic caused the radical suspension, for about two years, of major events related to the entertainment sphere, such as concerts and festivals, both outdoor and indoor. The opportunity for the reopening of such events, in a global context characterized by the real effects of climate change, requires the necessary adjustment to a sustainable approach—environmental, social, economic, and medical. Organizers of large events have the opportunity to recover their spaces and their audiences by also taking advantage of research that has matured in recent years; especially the possibilities offered by new technologies, which developed very quickly because they were stimulated by the requirements of the Pandemic. In order to contribute to the achievement of the goals of Agenda 2030, knowledge of, and a conscious desire to manage events in a sustainable manner is essential.

In planning and managing a sustainable event, there are many aspects to consider: the choice of location, the availability of infrastructure, the means of transportation, the procurement of materials for the construction of facilities, water and food supplies, as well as waste disposal and the restoration of venues to their original condition.

M. Mancini (✉)
University of Ferrara, Ferrara, Italy
e-mail: marco.mancini@unife.it



T. Walker et al. (eds.), *The Role of Design, Construction, and Real Estate in Advancing the Sustainable Development Goals*, Sustainable Development Goals Series, https://doi.org/10.1007/978-3-031-28739-8_10

As many fairs, concerts, and sporting events are often located outside urban centers, sustainability in transportation can be increased through car-sharing systems and the implementation of public transportation, with ad hoc rides managed through an app or reservation systems that can optimize the number of rides.

For access control, it may be advantageous to use new management technologies related to security, such as artificial intelligence systems (already in use at airports) to remotely check for temperature, presence of animals or illicit objects; or automatic QR code detection systems to check tickets, thus reducing the need for operators. Control logarithms could automatically open or close certain entrances based on incoming users and direct flows (as is presently done in strategic traffic-light management). The challenge is to use new technologies without negatively impacting the emotional component elicited by the event. Design choices (e.g., space and furniture design and choice of materials) must ensure safety without alienating participants.

There are many opportunities to properly manage waste: proposing strategic choices for catering, such as widespread drinking fountains so that bottles or water bottles can be filled, biodegradable or compostable packaging, and efficient, incentive-based collection systems.

The 4.0 and 5.0 paradigms can help in many aspects related to the management of large events. The opportunity to be seized upon involves harnessing these technologies in a lasting way, forming the basis for a lean, sustainable and resilient approach to the dramatic changes that have come to characterize this decade. This approach can generate positive results only if the actors involved in managing the event (site managers, product and service providers, communication managers) follow a well-defined plan, having clearly delineated tasks. Since these major events have both immediate and far-reaching impacts, it becomes possible to promote the global expansion of ethical environmental and sustainable choices.

In consideration of the issues pertaining to the 2030 agenda, a sustainable management of big events can contribute to the achievement of many goals: ensure access to water and sanitation for all (Goal 6, targets 6.4, 6.A, 6.B), ensure access to affordable, reliable, sustainable and modern energy (Goal 7, targets 7.1, 7.2, 7.A), promote inclusive and sustainable economic growth, employment and decent work for all (Goal 8, targets 8.2, 8.3, 8.4, 8.5, 8.8), make cities inclusive, safe, resilient and sustainable (Goal 11, targets 11.2, 11.3, 11.7, 11.A), ensure sustainable consumption and production patterns (Goal 12, targets 12.3, 12.5, 12.6), take urgent action to combat climate change and its impacts (Goal 13, targets 13.3), sustainably manage forests, combat desertification, halt and reverse land degradation, halt biodiversity loss (Goal 15, targets 15.2,. 15.5, 16.B), revitalize the global partnership for sustainable development (Goal 17, targets 17.3, 17.17).

## Big Events

The year 1851 is a pivotal year for all scholars of architecture, design, and artistic craftsmanship. It coincides with the first documented big event: the London World's Fair. On that occasion, the famous Crystal Palace was built, a building constructed with "dry" technology, typical of buildings such as garden greenhouses, made of iron and glass. If we consider the data (about 14,000 exhibitors and 6 million visitors) (Moretti, 2005), the design (specifically tailored for the event), the attention to the environment (centuries-old trees present in Hyde Park were not cut down but incorporated into the structure) and, finally, the subsequent disassembly and reuse of the site as a museum, we can reasonably conclude that the London World's Fair was the first major sustainable event in the modern history.

For a long period of time after that, universal expositions were major events that left their mark in terms of enhancing the places where they were held: consider the Eiffel Tower, built in 1889 for the Paris Expo, and the first Ferris wheel at the Chicago Expo in 1893. Although universal expositions involved a variety of novel architectural and urban planning areas, at different design scales, the real theme, common to all the major events is the temporal aspect. The term *event* in fact refers to something unaccustomed, temporary, thus including both the temporary use of removable artifacts (expos, large concerts, traveling events) and the temporary use of fixed artifacts (Venice biennale, fairgrounds, but also large installations for summer events such as the Bournemouth Pier).

In the case of an event featuring the temporary use associated with removable artifacts, it is clear that the problems related to a proper sustainable approach are greater: in a large festival, in addition to the main stage there will be smaller stages, dressing rooms, restrooms, ticket offices, bars, restaurants, kitchens, storage rooms, press rooms, lounge rooms, infirmary, etc., all of which require design, construction, assembly and disassembly. The first step in determining the sustainability of these events lies precisely in verifying the need for them. Without considering the induced activities generated by the catering, hospitality, and transportation sectors, the key question remains: is the major event aimed at enriching the host city (consumption of local resources with benefits for residents) or is the city itself being exploited as an attraction for paying spectators (consumption of local resources with benefits destined elsewhere)?

## Plastic Reduction

In Glastonbury (Somerset, UK), one of the largest music festivals in the world is organized, with about 200,000 spectators. Like almost all music festivals, the Glastonbury Festival takes place in the warm season, and in particular in the days around the Summer Solstice. There is a lot of dancing, a lot of sweating, and a lot of drinking. With reference to the 2014 edition, it was estimated by the organizers that one million empty plastic bottles were left on the ground (Townsend, 2014).

These numbers have raised great concern about both the actual environmental impact and the "Green" perception of the event. Many musicians, rock bands, and songwriters are at the forefront of environmental and social causes, and, for this reason, the Green festival image is also an important aspect to convey to spectators. In an attempt to curb the plastics problem, Glastonbury has banned the use of single-use plastic bottles and cups, wipes, and other non-biodegradable items since 2015, instead offering 37 recycled plastic kiosks from WaterAid with 20 units each to refill bottles at will. In addition, events supporting climate change awareness movements were planned among the side activities of the concerts. The presence of renowned British naturalist and science personality, David Attenborough, promoted the organizer's message of environmental awareness.

As a result of this, most large festivals today have programs to communicate sustainability, proposing, for example, composting and recycling, and, as far as possible, the total elimination of plastics. Glastonbury (UK), Bonnaroo (Tennessee, USA), Lollapalooza (Chicago, Illinois, USA) and Austin City Limits (Texas, USA) are among those that have given up plastic straws and cutlery, and the increasing presence of bottle-filling stations has significantly reduced the amount of waste. In 2018, 1.1 million bottles were saved at Lollapalooza, while at Bonnaroo, the Refill Revolution project prevented the consumption of more than 2 million plastic cups and bottles. The Plastic Pollution Coalition (PPC) has created a set of guidelines for plastic-free events that can be used by anyone, in stadiums or in small clubs. Glastonbury proved that eliminating plastic on a large scale is possible. In order to achieve this, however, a thorough examination of how water is distributed both backstage and to the audience is necessary. Transportation of liquids is the most important problem that needs to be solved, and, at Glastonbury, there are several tanks capable of holding approximately

800 thousand liters of water, supplied by Bristol Water via a pipeline. At Bonnaroo, the presence of infrastructure that allows filling stations to be supplied with water recovered from wells in the local area has been crucial. At urban festivals, such as Lollapalooza and Austin City Limits, water is sourced through infrastructure in the respective cities, while filtering takes place on site (Blistein, 2019).

Eliminating plastic and raising awareness of its proper use and disposal is probably the most easily communicated and pervasive aspect of environmental sustainability at events, but there are many other sources of environmental pollution, from $CO_2$ emissions to noise pollution.

How do large events, such as concerts and festivals, actually impact on the environment?

## Energy Consumption and $CO_2$ Emissions

The 350,000 spectators at the legendary three-day Woodstock festival left behind 1,400 tons of garbage in the mud at Winston Farm in Bethel, New York, 600 tons of which were still buried there in 1994, coinciding with the first big Woodstock anniversary (Spinelli, 2022).

In the way they are organized and in the way they are run, rock concert tours are undoubtedly a huge energy-consuming machine, even more than individual festivals and events. Consumption is highest for tours of the most famous rock stars, capable of filling stadiums, arenas, auditoriums with tens or hundreds of thousands of spectators. In a 2012 simulation (Landreman, 2012), an attempt was made to calculate the consumption of a concert by the British band Radiohead examining their performance at the Hollywood Bowl in Los Angeles, with a seating capacity of 18,000. Assuming two hours of performance, the sound system alone would consume 0.15 GJ,[1] and the lighting system 0.23 GJ. The audience, just in the act of traveling to the concert location, would consume as much as 4.2 TJ[2] of energy. According to Stentiford (2007), 75% of event emissions were produced by live performances alone, of which 43 percent were attributable

[1] The Giga joule (GJ) is equivalent to one billion joules. The Tera joule is equivalent to 1,000 GJ. Source: International System.

[2] (18,000 people) ÷ (3 people / car) × (120 miles / car) ÷ (20 miles / gallon) = 36,000 gallons. Or, in terms of joules, (36,000 gallons oil) × (3.1 × 107 J / L) ÷ (0.264 US gallons / L) = 4.2 TJ. Source: Landreman (2012).

to spectator transportation. From analysis of this data, the basic criteria to be monitored in concerts and routines are:

- $CO_2$ emissions from electrical consumption (amplification systems, lights, etc.).
- $CO_2$ emissions from transportation (equipment and spectators).
- Pollution from waste.
- Noise pollution.
- Food and beverage supply and consumption.

Attempts have been made on different occasions to propose carbon offsets,[3] that is, asking the public to pay for an increase on their ticket to be spent on activities designed to offset carbon emissions generated, for example, during the concerts or on travel to the event. A 2016 study (Connolly et al., 2016) assesses when and how carbon offsets can be applied effectively. For a medium-sized band, the energy consumption on a four-date tour is about 200–250 kg of $CO_2$ emitted into the atmosphere, considering only the audience. These data are quite variable depending on the size of the events, but they take into account all the variables for transportation and allow the offsets to be estimated. This system can only work if those who ask for offsets can demonstrate a positive outcome of the operation, otherwise there is a risk of the opposite effect, i.e., an unnecessary greenwashing operation with ticket increases to the detriment of spectators. There are cases of unsuccessful offset projects: a well-known British band failed in its attempt to plant mango trees in India to offset emissions from their concerts, while another British band, in the face of a 15 percent increase on tickets, never proved that it had planted new trees.

## Artists' Environmental Awareness

*Woodman! Spare that Tree!*, a milestone piece by George Morris and Henry Russell, dated 1837, can probably be considered the first environmentalist-driven song. Later, other artistic works in music were based on themes of denunciation

[3] A carbon offset is a reduction or removal of emissions of carbon dioxide or other greenhouse gasses made in order to offset emissions produced elsewhere. Offsets are measured in tons of carbon dioxide equivalent.

or awareness of the environmental cause, largely inspired by Rachel Carson's book *Silent Spring* (1962): Joni Mitchell's *Big Yellow Taxi* (1970), inspired by her trip to Hawaii to a location in danger of being covered by a parking lot; The Beatles' *Mother's Nature's Son* (1968), a tribute to the beauty and spirituality of nature; Marvin Gaye's *Mercy Mercy Me (The Ecology)* (1971); Jackson Browne's *Before the Deluge* (1974); and Bob Dylan's *Licence to Kill* (1983), where the license to kill is the license humans have given themselves over nature (Khan, 2013).

In recent years, some artists have worked on specific issues related to the climate crisis. Some examples are Father John Misty's *Things It Would Have Been Helpful To Know Before The Revolution* (2017) and Andrew Bird's *Manifest* (2019). There are also many instances of active participation, from Woody Guthrie to Pearl Jam and Bjork and Paul McCartney's movement for *Meat-free Mondays*. Among the most active artists in this topic is the band Radiohead, who, in addition to being member of the Friends of the Earth campaign, understands and addresses the environmental impact of tours: from biofuel-powered busses and trucks to zero plastic consumption. Tom Yorke, the leader of the band, has, in the past, stated that, for environmental reasons, he would prefer to stop live performances.

Another artist committed to these issues is Sheryl Crow, who claims to have saved about 750 tons of $CO_2$ on her 2010 tour alone, thanks to biodiesel, compostable catering, and various forms of recycling during tour stays. She partnered with car-sharing company (ZimRide) to reduce audience waste and also joined the campaign stopglobalwarming.org. Don Henley of the Eagles created and funded, with part of the income from his concerts and records, the Walden Woods Project and the Caddo Lake Institute, which researches ecology and works to preserve Texas's only natural lake. Pearl Jam, active on the environmental front for more than 20 years, donated $100,000 in 2006 to nine organizations working on renewable energy and climate change. Very prominent was their protest in 2010 over the Deepwater Horizon environmental disaster in the Gulf of Mexico. In the same year, the band Green Day partnered with the U.S. Natural Resources Defense Council for the *Move America beyond Oil* campaign. The Dave Matthews Band, in addition to having donated more than $8 million to various charities over the years, has been cultivating a project since 2009 to get their fans involved. Fans receive free passes to concerts if they decide to engage in practical green initiatives that are promoted on the sidelines of their concerts, as well as being urged to carpool to reduce fuel consumption (Milano, 2019).

## Best Practice and Regulations for "the Sustainable Event"

Artists play an important social role in sustainability. Thanks to their popularity they can transmit messages that gradually are adopted by the public. As the years passed, it was finally possible to transform best practices into guidelines and standards.

### The Sustainable Events Guide

A sustainable event is one designed, organized and implemented in a way that minimizes potential negative impacts and leaves a beneficial legacy for the host community and all involved.[4]

The Sustainable Events Guide, proposed by the United Nations Environmental Programme, UNEP (2012), is aimed at offering organizers a wealth of concrete, easily understandable and accessible advice including guidance on management issues, sector-specific recommendations, and action-oriented checklists. It consists of six sections:

Section 1: Sustainable events as an opportunity for change—an introduction to the concept of sustainable events and the benefits these can bring to event organizers and other engaged stakeholders.

Section 2: Managing and communicating sustainable events—guidance on management and communication aspects of sustainable events, with a special focus on the engagement of relevant stakeholders.

Section 3: Implementing sustainable events—a summary of the main conference areas and the actions that can be taken to reduce potential negative impacts and increase benefits (with a special focus on venue selection, marketing and communication, accommodation, transport, and catering). This section also covers recommendations on how to embed social criteria throughout event preparation and implementation (small local business support, social integration, food waste, etc.).

Section 4: Climate-neutral and climate-friendly events—an overview on the topic of carbon offsets and proposals for calculating and offsetting the remaining greenhouse gas emissions generated by an event.

---

[4] Adapted from the Green Meeting Guide 2009 and based on the principles developed at the ICLEI. Greening Events Symposium in Barcelona, Spain, September 2004.

Section 5: Reporting on sustainable events—guidelines on how to report on an event's sustainability measures.
Section 6: Sustainable Events Checklists—detailed sustainable recommendations for the day-to-day organization of an event (UNEP, 2012, pp. 8, 9).

The guide to sustainable events is undoubtedly a valuable source of advice and best practices for organizers who, however, can enshrine the sustainability of their event by adopting the international ISO 20121 standard.

## The ISO 20121 International Standard

The international standard, ISO 20121:2012 defines the requirements for a management system aimed at designing and implementing environmentally, socially, and economically sustainable events. Developed by the ISO/PC 250 Project Committee "Sustainability in Event Management", it involved more than 30 countries as participating members or observer members, and was first applied in the 2012 London Olympics. It was reviewed and confirmed in 2017 by the International Standard Organization (therefore the 2012 version still remains current).

As any management system structured according to the HLS Model,[5] it starts from the analysis of the context. For the definition of the external context, factors related to the surrounding environment that may affect the organization are considered, such as the cultural, social, political, technological environment not only nationally and internationally, but also regionally or locally. As for the internal context (the internal condition of the organization), it is required to consider the structure of the organization, the policies implemented to achieve certain objectives, the corporate culture, and applicable standards and guidelines.

Once the context has been defined, the next step is to determine the stakeholders involved in the delivery of the event, and their needs and expectations. The standard provides for the preparation, implementation, and maintenance of a procedure by which stakeholders are identified, roles defined, and the joint commitment on the aspects of sustainable development (Environmental, Social and Economic) are determined. These include, for example, the workforce that will be in place during the course of the event, the identification of suppliers, the census data of participants, regulatory bodies involved, and, finally, the description

[5] An HLS (an acronym for High Level Structure) is a structure common to all new ISO standards in order to achieve the best interaction between multiple integrated management systems.

of the community that is affected by the delivery of the event, enunciating what benefits they may gain from hosting the event.

During the planning stage, the organization must analyze risks and opportunities to ensure that the sustainable event management system can achieve the intended outcomes, predict and reduce undesirable effects, and achieve continuous improvement. Leadership must ensure that sustainable development policies and objectives are defined, including providing the resources needed to achieve them (both human and material), as well as monitor the performance of organized events in the process of continuous improvement.

UNI ISO 20121 requires the implementation of a procedure, documented and shared with stakeholders, in which the aspects of sustainable development are indicated:

- Environmental: resource conservation, reduction of atmospheric emissions, biodiversity, and nature conservation.
- Social: protection of health and safety in the workplace, equality between men and women, and rights of local people.
- Economic: Innovation and Fair Trade.

Once the various aspects have been defined, it is necessary to identify and determine the legal requirements applicable to the organizational context and then identify the objectives to be achieved, which will have to be monitored, communicated, and constantly updated.

In order to achieve the established objectives, it is essential that

- Resources (material and human) are sufficient and adequate.
- People are competent and aware of their role in the organization.

The organization must monitor the performance and objectives of the management system at planned intervals, reviewing it periodically and reporting and resolving nonconformities.

The benefits of implementing the UNI ISO 20121:2012 standard are many, both for organizers and the environment, and relate to improving event management and supply chain impacts, increasing staff awareness of sustainable issues, and ensuring compliance with environmental regulations (Berni & Tutolo, 2022).

## Crucial Aspects for "Sustainable Event" Planning

In this section, we summarize the most crucial aspects for the design of a sustainable event, following a systemic approach (Bistagnino, 2009) based on our research and from direct interviews with event organizers.

### Location Selection

Location selection is the first step in the whole process of defining the sustainability of an event, because this is where the main motivation for the event and its connection to the surrounding area originates. The presence of infrastructures such as roads, railways, as well as parking or accommodation facilities, water supply, and sewage collection facilities are essential.

### Temporary Installations

The possibility of staging a temporary event in a location free of permanent structures is posed in relation to the type of staging, the temporary constructions that are built, the choice of materials, the arrangement of the construction phases and especially the post-use phase, whereby natural materials such as wood, or steel structures can be reused. In some cases, sustainable (both environmentally and economically) toilets have been proposed, for example, compost toilets with collection and reuse from fertilizer companies.

In some circumstances, ad hoc arrangements have been made, for example, the recent installation of the Abba Arena (Abba Voyage, 2022). It is considered the largest dismountable Arena in the world, made of steel and wood, installed in London and planned to be in use for 5 years after installation. After this period, it will be dismantled and moved elsewhere. The building, designed by STUFISH Entertainment Architects, can accommodate up to 3,000 people. The advantages of this type of installation are many: proximity to local public transport (so reaching the location does not require the use of private vehicles, thus avoiding a major component of $CO_2$ emissions); a design specifically targeted to the type of event (so noise and light pollution are contained within, thanks to the type of architecture and technology used); and the possibility of changing the configuration to adapt to the requirements of the next location where the arena will be installed.

## Energy Consumption

Facilities that provide energy from renewable sources are good starting points, but in addition to energy generation, it is essential to analyze the routes and means of transport used, which, as we have noted, have very high values in terms of environmental impact. Some examples of attempts to reduce consumption are the use of solar energy (as in the case of Terraforma festival through Etica Sgr and the start-up Liter of Light, both from Italy), geothermal energy (as in the site-specific case of the Secret Solstice Festival in Reykjavik, Iceland), and conscious user involvement (such as Ciclo Cinema, in Italy, where exercise bikes power outdoor cinemas).

## Catering

In the area of catering, some conflicting observations can be made. On the one hand, is the necessary attention to the quality of the food offered: short-chain crops, sustainable agriculture, organic food, ethical employment, and other similar considerations (Petrini, 2019). On another side is the diversification of offers proposed to meet the needs of participants: a variety of menu choices, such as vegetarian, vegan, celiac, children's menu, fish menu, local or ethnic cuisine.

The greater the diversification, the greater also are the number of vehicles that must be utilized to move the equipment for each event. Many vans or kiosks that offer many different types of food have a huge impact in terms of $CO_2$ emissions, despite the fact that the products sold are defined as "Green". Therefore, a lot of thought is required, especially in order to avoid greenwashing practices. One, still unresolved, issue is the management of food that, not consumed, could be reused for redistribution to people in need. The solution brought to the 2019 DGTL Festival in Amsterdam pushes beyond the zero-kilometer solution by partnering with the Instock restaurant, known for basing its menu on food waste. Produce is collected from a variety of vendors with surplus food and beverages that should not be sold as damaged goods from supermarkets and markets. Instock chefs are able to create dishes without first knowing what will be available. DGTL's media campaign highlights how menus have not been chosen specifically for the target audience, but have to adapt according to the local surplus.

## Waste Management

Proper waste management is the sustainable aspect most directly perceived by users and therefore most often communicated. It was precisely from waste management, especially disposable plastic bottles that the so-called "green" turn of major events started around 2015. This was followed by legal obligations, such as the European (EU) Directive (EU) 2019/904, Single Use Plastic (SUP) banning disposable items: straws, q-tips, plates and cutlery, cocktail stirrers, balloon sticks, Styrofoam food and beverage containers, just to name a few.

## Mobility

Events organizers have two types of mobility to manage: internal mobility, which varies according to the size of the venue, and external mobility, which concerns users access to the event. From this point of view, many events are moving forward by guaranteeing subsidized rates for public transportation or by incentivizing vehicle sharing. This aspect, as already mentioned, is subservient to the primary one, inherent in the choice of location.

## Safety and Security

The issue of safety is particularly important because sometimes safety of places, equipments, and activities carried out do not fall within specific and determined legislation but through voluntary declarations that rely on the assessment of the declarant.

Safety is above all connected to structural factors (building, stage, and equipment safety) and traffic (access routes, room capacity, escape routes). An important issue is the regulation and separation of access routes during all the phases of the event.

The recent spread of the COVID-19 virus has brought to light a series of problems related to crowds and disease transmission. For this reason, during the pandemic almost all activities related to live performances have been curtailed and venues have been closed. Health and sanitary safety concern, after the resolution of the pandemic, continue to be included along with other aspects relating to safety (static safety, fire prevention, escape routes, maintenance of structures, and equipment).

Any public entertainment event provides different areas, with access allowed to different categories of people (artists, technicians, management, journalists, photographers, the public). In order to effectively manage access, to enable or disable entrances, or to monitor movement, geolocation, and tracking technologies could be used, such as GPS integrated into smartphones, or RFID bracelets. If we consider the event location as a place potentially at risk to both terrorist attacks and the spread of disease, then the use of Artificial Intelligence (AI) could help to reduce the risk. For example, AI makes it possible to generate an alert when large backpacks are recognized (which may potentially contain explosives), or to characterize movements attributable to people in an altered state of consciousness (alcohol, drugs). AI could also help to recognize specific categories of people, such as disabled people in wheelchairs, and automatically open the turnstiles dedicated to them without the need of staff. Another use could be the control of vaccination certificates and temperature monitoring (i.e. to allow access only to certain categories). Where there are specific areas dedicated to the public with pets, AI could recognize animals and enable owners to selectively access only those designated areas.

## The Contribution of Design

The design sector, in its broadest extent, can contribute in many ways to the implementation of sustainable events, with a view to achieving the goals of Agenda 2030. The following are some examples of service and product design proposals:

- Advanced integrated systems for organizing sustainable events, such as new booking applications, and new ways to manage reimbursements for canceled events.
- Specific instruments for the tracking, control, and management of entrances to events, through the use of new sensors and IT technologies implemented ad hoc, with the definition of areas based on the assigned roles, time of use, sanitary passports, etc.
- Communication of smart and sustainable car-sharing system or other solutions (i.e. the activation of special public transport vehicles) for the reduction of pollution and to reduce the number of private vehicles.

- Design of installations with advanced integrated features (holograms, video installations, multisensory experiences, visual/auditory/olfactory associations, etc.) in order to optimize communication, to be used before, during and after the event.
- Design of podcasts, videos, or virtual assistants where this could help to avoid unnecessary contact between people, in any indoor or outdoor location.
- Coordinated and integrated websites with interactive experiences, for use before, during, and after the event. The permanence of websites is fundamental to continuous improvement, communicating best practices, and active action for reaching the Agenda 2030 goals.

## The 2030 Agenda

In consideration of the issues pertaining to the 2030 agenda, we highlight how sustainable management of major events, especially those in the music industry, can contribute to the achievement of the agenda objectives, including specific goals and targets.

### Goal 6: Ensure Access to Water and Sanitation for All

In major music-related events, the issue of water sustainability has been addressed, and many successful initiatives has been implemented related to optimizing flows, supplies, recycling and reuse that demonstrate achievement of these goals is possible.

The targets of this goal that can be achieved are:

6.4 By 2030, substantially increase water-use efficiency across all sectors and ensure sustainable withdrawals and supply of freshwater to address water scarcity and substantially reduce the number of people suffering from water scarcity.
6.A By 2030, expand international cooperation and capacity-building support to developing countries in water- and sanitation-related activities and programs, including water harvesting, desalination, water efficiency, wastewater treatment, recycling, and reuse technologies.
6.B Support and strengthen the participation of local communities in improving water and sanitation management.

## Goal 7: Ensure Access to Affordable, Reliable, Sustainable, and Modern Energy

Energy-sustainable events were implemented through the installation of off-grid platforms and solar panels aimed at optimizing the energy resources used, thus demonstrating that both the technologies and resources exist to achieve the following targets:

> 7.1 By 2030, ensure universal access to affordable, reliable and modern energy services.
> 7.2 By 2030, increase substantially the share of renewable energy in the global energy mix.
> 7.A By 2030, enhance international cooperation to facilitate access to clean energy research and technology, including renewable energy, energy efficiency, and advanced and cleaner fossil-fuel technology, and promote investment in energy infrastructure and clean energy technology.

## Goal 8: Promote Inclusive and Sustainable Economic Growth, Employment, and Decent Work for All

In both *The Sustainable Events Guide* and *ISO 20121* specifications, the theme of social sustainability is particularly highlighted: in addition to health and safety in the workplace, respect for equality between men and women as well as respect for the rights of local people is called for. Many of the objectives also concern staff awareness of sustainable issues and ensuring compliance with environmental regulations. All of this fits positively into efforts to achieve the following targets:

> 8.2 Achieve higher levels of economic productivity through diversification, technological upgrading, and innovation, including through a focus on high-value added and labor-intensive sectors.
> 8.3 Promote development-oriented policies that support productive activities, decent job creation, entrepreneurship, creativity, and innovation, and encourage the formalization and growth of micro-, small- and medium-sized enterprises, including through access to financial services.

8.4 Improve progressively, through 2030, global resource efficiency in consumption and production and endeavor to decouple economic growth from environmental degradation, in accordance with the 10-year framework of programs on sustainable consumption and production, with developed countries taking the lead.
8.5 By 2030, achieve full and productive employment and decent work for all women and men, including for young people and persons with disabilities, and equal pay for work of equal value.
8.8 Protect labor rights and promote safe and secure working environments for all workers, including migrant workers, in particular women migrants, and those in precarious employment.

## Goal 11: Make Cities Inclusive, Safe, Resilient, and Sustainable

Many major events are designed to take advantage of existing local infrastructures such as active public transport systems. The proceeds from the increased use of public transport system can then be used to improve existing infrastructure, with positive impacts on the resident population. This approach also seeks to avoid the additional $CO_2$ emissions caused by the use of private vehicles. The sustainable management of major events can help the achievement of the following targets:

11.2 By 2030, provide access to safe, affordable, accessible, and sustainable transport systems for all, improving road safety, notably by expanding public transport, with special attention to the needs of those in vulnerable situations, women, children, persons with disabilities, and older persons.
11.3 By 2030, enhance inclusive and sustainable urbanization and capacity for participatory, integrated, and sustainable human settlement planning and management in all countries.
11.7 By 2030, provide universal access to safe, inclusive and accessible, green and public spaces, in particular for women and children, older persons and persons with disabilities.
11.A Support positive economic, social, and environmental links between urban, peri-urban, and rural areas by strengthening national and regional development planning.

## Goal 12: Ensure Sustainable Consumption and Production Patterns

The most important role that can be played within large music events relates to both the reduction of food losses (see examples cited above) and the optimization of the waste cycle up to its reuse (for example, as fertilizer). The ISO 20121 standard is designed to encourage not only the communication of sustainable matrix activities, but also their monitoring in order to continuously improve them. The sustainable management of major events can help the achievement of the targets:

> 12.3 By 2030, halve per capita global food waste at the retail and consumer levels and reduce food losses along production and supply chains, including post-harvest losses.
> 12.5 By 2030, substantially reduce waste generation through prevention, reduction, recycling, and reuse.
> 12.6 Encourage companies, especially large and transnational companies, to adopt sustainable practices and to integrate sustainability information into their reporting cycle.

## Goal 13: Take Urgent Action to Combat Climate Change and Its Impacts

In addition to the initiatives already implemented and realized with the direct goal of reducing $CO_2$ emissions (mainly due to spectators transportation), major events have a greater social responsibility: informing and raising awareness about climate change and sustainability. Celebrities and artists (international, national and local) can convey messages aimed at engaging people. The sustainable management of major events can help the achievement of the following target:

> 13.3 Improve education, awareness-raising, and human and institutional capacity on climate change mitigation, adaptation, impact reduction, and early warning.

## Goal 15: Sustainably Manage Forests, Combat Desertification, Halt and Reverse Land Degradation, Halt Biodiversity Loss

In big events, these goals often take the form of carbon-offset proposals in which artists pledge to plant trees, especially in areas at risk of desertification, in order to offset the $CO_2$ emissions generated by their tours. The sustainable management of major events can help the achievement of the following targets:

15.2 By 2020, promote the implementation of sustainable management of all types of forests, halt deforestation, restore degraded forests and substantially increase afforestation and reforestation globally.
15.5 Take urgent and significant action to reduce the degradation of natural habitats, halt the loss of biodiversity and, by 2020, protect and prevent the extinction of threatened species.
15.B Mobilize significant resources from all sources and at all levels to finance sustainable forest management and provide adequate incentives to developing countries to advance such management, including for conservation and reforestation.

## Goal 17: Revitalize the Global Partnership for Sustainable Development

Large events almost always require shared goals between the event managers and the host governments. These partnerships have generated significant induced and economic returns for residents. Often artists have set out to raise awareness of sustainability issues. *Live Aid*, held in 1985, simultaneously in London and Philadelphia, was organized explicitly to raise funds for the Ethiopian famine, and far exceeded initial expectations. The widespread media coverage of such events is an important step in favor of raising awareness of the environmental and social issues.

The sustainable management of major events can help the achievement of the following targets:

17.3 Mobilize additional financial resources for developing countries from multiple sources.

17.17 Encourage and promote effective public, public–private, and civil society partnerships, building on the experience and resourcing strategies of partnerships.

## Conclusion

What has been covered in this chapter should help clarify a fundamental point: in a large event, the responsibility to act sustainably is shared between organizers and spectators. In the choice of location, the type of transportation used, the conscious consumption of primary goods such as water and food, the disposal or recovery of waste, there is a continuity. A sense of community is created through the sharing of a unique experience, of listening to "our song" with friends and fellow attendees and also with other fans, people until then unknown. The positive memory of such events lingers, for months or even years following the event itself.

This association between environmental, social, and ethical behavior and the powerful emotion and memory created by a mass concert, festival, or event is a unique opportunity for promoting sustainable practices. This is the strength of great musical events. To intelligently, sustainably, and consciously exploit this opportunity lies an interesting and, at the same time, promising challenge for the future of our planet and our society.

**Acknowledgements** The author would like to express his thanks to Eleonora Trivellin for her valuable input and advice with reference to the topics covered in this chapter.

## References

Abba Voyage. (2022). Retrieved July 15, 2022, from https://abbavoyage.com/thearena/

Berni, A., & Tutolo, U. (2022). *Sostenibilità degli eventi: come e dove agire*. Retrieved July 27, 2022, from https://www.esg360.it/normative-e-compliance/sostenibilita-degli-eventi-come-e-dove-intervenire/

Bistagnino, L. (2009). *Design sistemico*. Bra: Slow Food.

Blistein, J. (2019). *Come Glastonbury guiderà la rivoluzione ambientalista*. Retrieved July 27, 2022 from https://www.rollingstone.it/musica/come-glastonbury-guidera-la-rivoluzione-ambientalista/450742/

Connolly, M., Dupras, J., & Séguin, C. (2016). An economic perspective on rock concerts and climate change: Should carbon offsets compensating emissions be included in the ticket price? *Journal of Cultural Economics, 40*(1), 101–126.

International Standard Organization (ISO). (2017). *ISO 20121:2012 Event sustainability management systems—Requirements with guidance for use*. ISO/TMBG Technical Management Board–groups. Reviewed and confirmed in 2017.

Khan, R. (2013). Environmental activism in music. In J. Edmondson (Ed.), *Music in American Life: The songs, stories, styles, and stars that shaped our culture* (pp. 412–417). Greenwood Press.

Landreman, P. (2012). *Killer cars: The energy impact of a radiohead concert* (Work for Physics of Energy course, Stanford University). Retrieved July 14, 2022, from http://large.stanford.edu/courses/2012/ph240/landreman2/

Milano, M. (2019). *I concerti possono essere sostenibili?*. Retrieved July 10, 2022, from https://oggiscienza.it/2019/07/31/concerti-sostenibili-ambiente/index.html

Moretti, S. (2005). *Esposizioni universali*. Retrieved July 28, 2022, from https://www.treccani.it/enciclopedia/esposizioni-universali_%28Enciclopedia-dei-ragazzi%29/

Petrini, C. (2019). *Buono, pulito e giusto*. Slow Food.

Spinelli, A. (2022). *Pace, amore e rifiuti da Woodstock a noi*. Retrieved June 28, 2022, from https://www.quotidiano.net/economia/impreseitaliane/pace-amore-e-rifiuti-da-woodstock-a-noi-1.7486704

Stentiford, C. (2007). *Ecological footprint & carbon audit of radiohead North American tours*. Best Foot Forward Ltd.

Townsend, M. (2014). *Glastonbury goes green: Festival declares war on plastic water bottles*. Retrieved June 20, 2022, from https://www.theguardian.com/music/2014/jun/22/glastonbury-declares-war-plastic-water-bottles

United Nations Environmental Programme (UNEP). (2012). *Sustainable events guide, give your large event a small footprint*. Retrieved June 10, 2022, from https://uist.acm.org/uist2019/sustainability/SustainableEventsGuideMay302012FINAL.pdf

# Equity in the Built Environment in Least Developed Countries: The Case of Rural Municipalities in Nepal

Robert Brian Smith

## Introduction

Nepal is a least developed country in the Himalayan Mountain range north of India. Its development is hampered by a lack of easy accessibility due to the rugged terrain. This chapter discusses the challenges of accessibility for the development of rural municipalities as they struggle to meet the sustainable development goals.

## The Challenges

Many functions have been devolved to local authorities under the Constitution of Nepal (2015). These include management of the local services, local development plans, and projects; basic[1] and secondary education; basic health and sanitation; local market management, environment protection, and biodiversity; local roads, rural roads, agricultural roads, and irrigation. A discussion of the decision to designate/devolve these functions to local authorities, given their financial

[1] In the Nepal education system basic level education is provided in grades 1 to 8 (Gurung, 2021). Basic level education would becalled primary or elementary in other systems.

R. B. Smith (✉)
School of Law, University of New England, Armidale, Australia
e-mail: robert@aecconsultants.asia

 
T. Walker et al. (eds.), *The Role of Design, Construction, and Real Estate in Advancing the Sustainable Development Goals*, Sustainable Development Goals Series, https://doi.org/10.1007/978-3-031-28739-8_11

insecurity and capacity limitations, is necessary but outside the scope of this chapter.[2]

As will be seen, Nepal faces significant challenges, especially as it prepares to graduate from its status as a least developed country following a five-year transition period (United Nations General Assembly, 2021). At its triennial review in 2024, the Economic and Social Council will determine whether a further transition period is required.

## Topography and Geology

About 77% of Nepal's area lies in the hill and mountain regions (Mulmi, 2009, p. 149). It is relatively small, with an area of 147,181 square kilometers, a north–south length of up to 200 km, and an east–west length of up to 800 km. Nepal has complex topography and geology (Dahal, 2010; Upreti, 1999) and is in a seismically active zone (Dey, 2015; National Planning Commission, 2015; 'Nepal Earthquake Case Studies: Engineering, Infrastructure, Design', 2015), which activates landslides (Valagussa et al., 2021); and has a summer monsoon season (Poudel, 2022) and causes flooding as well as landslides (Bell et al., 2021; Dahal, 2010; Muñoz-Torrero Manchado et al., 2021; Shrestha & Mishra, 2020).

The complexity of the topography and geology can be gaged from the major physiographic units across Nepal from the south to the north (Upreti, 1999).[3] The major tectonic zones of Nepal are:

a. Terai Zone—northern edge of the alluvial Indo-Gangetic basin;
b. Churia (Siwalik equivalent) Zone—sedimentary basin deposits;
c. Lesser Himalayan Zone—complex geology including schist, phyllite, gneiss, quartzite, granite, limestone, and marble rock types;
d. Higher Himalayan Zone—Precambrian gneisses, schists, and marbles; and
e. Tibetan-Tethys Zone—limestones, shales, sandstone, and other sedimentary deposits.

[2] It has been discussed extensively see for instance: Acharya (2018), Acharya & Zafarullah (2020), Adhikari & Upadhyaya (2020), Chaudhary (2019), Paudel & Sapkota (2018), Pradhan (2019).

[3] The author has restricted this discussion to the variety of physiography units rather than providing a detailed geological and geomorphological discussion.

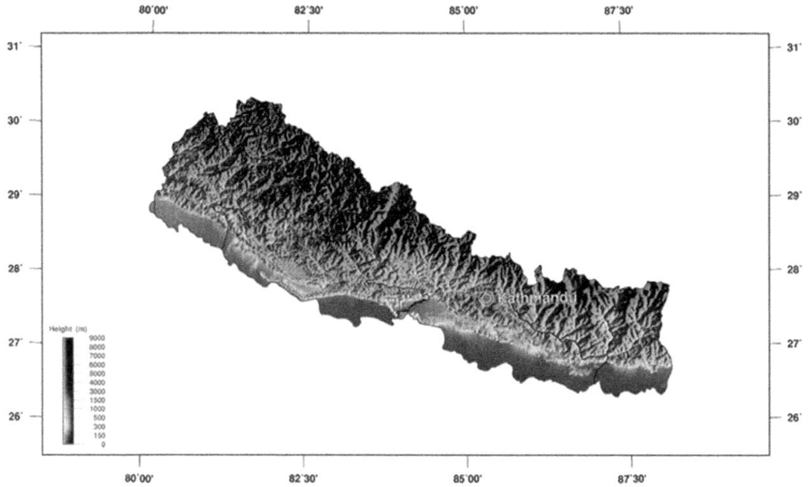


**Fig. 1** Topographic map of Nepal (*Source* https://www.mapsland.com, Creative Commons Attribution—ShareAlike 3.0 Licence)

The topography is shown in Fig. 1.

A devastating earthquake measuring 7.9 on the Richter scale occurred in Nepal on 25 April 2015 (Dey, 2015, p. 28). A further earthquake struck on 12 May 2015 (National Planning Commission, 2015, p. iii). The major earthquake took the name of the area west of Kathmandu where it occurred, namely Gorkha. As a result of these two earthquakes, close to 9,000 lives were lost, and around half a million houses were destroyed (National Planning Commission, 2015). To put this in perspective, it was estimated at the time that the lives of eight million people, almost one-third of the country's population, were affected.

> The destruction was widespread covering residential and government buildings, heritage sites, schools and health posts, rural roads, bridges, water supply systems, agricultural land, trekking routes, hydropower plants and sports facilities. The geodetic network centres including horizontal and vertical control points have been damaged in a manner that will affect reconstruction planning. Rural areas in the central and western regions were particularly devastated and further isolated due to road damage and obstructions. In the worst hit areas, entire settlements, including popular tourist destinations like Langtang, were swept away by landslides and avalanches triggered by the earthquakes. Due to the weakened, ruptured, and destabilized slopes and surfaces, the vulnerable areas have now become even more susceptible to flooding and landslides that can occur during the monsoon. (National Planning Commission, 2015, p. xi)

**Table 1** Classification of local government units based on infrastructure and social development status [adapted from Samiti (2019) and Wikiwand (2021)]

| Classification | Type | Local Government Unit | |
|---|---|---|---|
| | | Municipality | Rural Municipality |
| Grade 'A' | Very remote | 17 | 145 |
| Grade 'B' | Remote | 44 | 174 |
| Grade 'C' | Fairly accessible | 134 | 141 |
| Grade 'D' | Accessible | 98 | 0 |
| Total | | 293 | 460 |

Poorer rural areas were affected more than towns and cities due to the poorer quality of housing. "More women and girls died than men and boys, partly because of gendered roles that disproportionately assign indoor chores to women" (p. xii). Seven thousand schools were completely or significantly damaged but as Saturday is the weekly holiday schools were closed.

The World Bank estimates that the rural population is around 23 million, over 79% of the total population of Nepal (World Bank, 2022). The largest rural municipality (Gaunpalika in Nepali) has a population of around 60,000, with the smallest being Manang's Narpa Bhumi Rural Municipality, which has only 442 people (Kharbar, 2022).[4] As shown in Table 1, 61 municipalities (towns) and 319 rural municipalities are in remote or very remote areas of the country. Even those that are relatively accessible can have difficult access as they are often in hilly and mountainous regions.[5]

All local government units within Nepal are divided into wards that provide essential services to other population centers, such as villages. The wards require access to each municipality's administrative center, and the municipality must be connected to the outside world through reliable mobile and high-speed internet network connections. However, many centers lack easy access to services. Thus, a prioritization program must be developed to ensure equity.

[4] Narpa Bhumi Rural Municipality is located in Mustang, an isolated region in northwestern Nepal in the rain shadow of the Himalayan Mountain Range and is on the ancient Silk Road between China and India.

[5] For instance, Madi Rural Municipality is approximately 34 km from Pokhara, a major regional city but the trip takes around 1 h 45 m by light vehicle and even longer by bus due to the hilly/mountainous terrain and the need to descend to cross the Madi River and then ascend again to Thumako Danda, where the Municipality is based.

## Development in Rural Nepal

As can be seen in Table 2, the rural population is estimated to be over 23 million and increasing at a rate of around 1.3% as of 2020. While there have been continuing improvements in most development statistics from 2018–2020, the percentage of the rural population using safely managed drinking water services has dropped from 19.0% in 2018 to 15.7% in 2020.

In these rural municipalities, access to facilities is defined as "rural access." The most advanced measure is the Rural Access Index (RAI). This technology-based measure has been developed to determine the number of persons in rural areas that live within 2 km of a road in good or fair condition (Purdie et al., 2016). When the most recent RAI was reported for Nepal in 2016, it was found that 54% of the rural population lived within 2 km of such roads. That meant that in 2015, nearly 10.3 million residents did not have easy access to goods and services. In the plains of the Terai, the RAI reached 80% in some districts due to the combination of high population density and high road density. In the more rugged northern regions, the road density is much lower, and the RAI in many places was less than 20% (Purdie et al., 2016).

**Table 2** Selected development statistics of rural population in Nepal [extracted from World Bank Databank (2022)]

| Parameter | 2018 | 2019 | 2020 |
|---|---|---|---|
| People practicing open defecation, rural (% of rural population) | 17.8 | 14.5 | 11.2 |
| People using at least basic drinking water services, rural (% of rural population) | 89.0 | 89.6 | 90.2 |
| People using at least basic sanitation services, rural (% of rural population) | 68.7 | 72.6 | 76.7 |
| People using safely managed drinking water services, rural (% of rural population) | 19.0 | 17.4 | 15.7 |
| People using safely managed sanitation services (% of population) | 43.3 | 45.5 | 48.6 |
| People with basic handwashing facilities, including soap and water, rural (% of rural population) | 55.6 | 57.2 | 58.8 |
| People practicing open defecation, rural (% of rural population) | 17.8 | 14.5 | 11.2 |
| Rural population (million) | 22.54 | 22.84 | 23.14 |
| Rural population growth (annual %) | 1.2 | 1.3 | 1.3 |

**Fig. 2** Typical house on the Terai (plains)

A selection of photographs[6] depicting development in various areas of Nepal is shown in Figs. 2, 3, 4, 5, 6 and 7. Landslides are an annual occurrence, although highly destructive landslides such as those shown in Fig. 8 and Fig. 9 are less common. Figure 8 shows the extent of the landslide that blocked the Araniko Highway, the main route from Kathmandu to Tibet. The Highway is subject to regular landslides (Cowan, 2021). Bahrabise on the east bank of the Koshi River along the Araniko Highway needs protection from the fast-flowing river and possible local landslides (Fig. 10). Finally, the aerial view of typical development in the mountainous areas of Nepal with buildings on the ridges and access roads zigzagging from the ridge to the valley floor (Fig. 11). Depending on the locations of such access roads, they can be subject to minor water-induced slope failures as a result of major landslides from above or below the road.

[6] All photographs were taken by the author.

**Fig. 3** Brick houses in the hills southwest of Kathmandu

**Fig. 4** Access road to Thumako Danda

**Fig. 5** Houses at Thumako Danda

Malalgoda et al. (2014) investigated the challenges in creating a disaster-resilient built environment in Sri Lanka and identified the following challenges:

a. Regulatory frameworks were inadequate to reduce disaster impacts and improve resilience.
b. Cities are often unplanned,[7] and there is rapid urbanization.[8]
c. The building stock is often old, and the infrastructure is far from resilient.
d. Many structures are unauthorized and lack basic facilities.
e. Institutional arrangements are often fractured.
f. While municipalities are responsible for building approvals, they often lack the capacity in both human resources and tools to undertake the role.

[7] This also applies to towns and villages in Nepal. which are often unplanned.

[8] In Nepal the rural population drifts to the larger population centers, particularly to the three co-joined cities of the Kathmandu Valley.

**Fig. 6** Police post at Thumako Danda

g. There is a lack of funding to ensure that newly built assets are hazard resilient and existing buildings are retrofitted to become hazard resistant.
h. Corruption and unlawful activities (pp. 739–741).

To a greater or lesser extent, all these challenges apply to rural municipalities in Nepal, especially following the devolution of power from the central government to the provinces and local units, with each having defined roles under the constitution (see, for instance: Smith and Smith, 2022). Recent studies have ascertained residents' views concerning their development priorities. Sapkota (2018) collected household survey data from three remote development committees in the hilly/mountainous district of Sindhupalchok from February to March 2014. The road network was minimal due to the "extremely rugged terrain" (pp. 14–15) This meant that access to services was impacted by the travel time required for pedestrian or animal access including basic river crossings. As a result, access to health services and education facilities were acknowledged as the most critical factors for their well-being. However, when asked about their priorities for infrastructure development, respondents gave access to roads and drinking water as their top priorities (p. 15). He considers that this resulted from them knowing that roads

**Fig. 7** The village aerial view from the access road to Thumako Danda

and water "are the key means to achieve their health and educational objectives" (p. 15). Access to schooling has been suggested to significantly impact human capital, especially in rural areas, as rural communities have considerable deficiencies relative to urban communities (Devkota et al., 2021, p. 12). Better access to education should reduce this deficiency. The difference between rural and urban capital was considered to be related to household income and occupation type in the household. Devkota et al. (2021) claim that the results of their study "reinforce the claim that an improvement in schooling access and road infrastructure is necessary, particularly in the vast rural sector of Nepal, if human capital development is to provide a greater contribution to national welfare" (p. 12).

Despite writing over a decade ago, Suvedi (2010) proffered a cautionary message, highlighting an issue that continues to be a potential flaw in design development projects, thereby affecting their sustainability. He noted that:

> Despite the fact that drinking water is a real need of the people of Hamsapur, the local capacity to maintain the system was missing. The project did not include training of the local people as electricians and plumbers. [. . .]. Instead of building local capacity to prevent possible problems in the operation of the drinking water supply, the project

**Fig. 8** Scar from the August 2014 landslide south of Bahrabise, which killed 15 people

> depended on outside expertise to repair the broken pumps. This increased the community's dependency on the outside world and reduced local capacity for development. (p. 193)

Road access is the key to achieving many of the SDGs, as is the provision of sustainable buildings. As a result, the focus will shift to two core infrastructure asset types: residential buildings, and roads.

## Resilient Residential Buildings

Modeling carried out more than three years before the Gorkha Earthquake by Ahmad et al. (2012) has shown that confined masonry construction schemes can avoid total structural collapse even with strong ground motion. In the confined masonry concept, the masonry is placed first and then confined within a reinforced concrete frame. Figure 12 shows a recent example of confined masonry as used in Cyprus, another earthquake-prone area.

**Fig. 9** Property destroyed by the August 2014 landslide

Confined masonry should ensure the safety of the occupants. In the case of large earthquake events, significant damage will require attention, such as cement grouting and/or the reinforced plastering of masonry walls. In addition, the confining elements may need repair. Unsurprisingly, they found that good-quality masonry material fares significantly better than low-quality material. They concluded that good-quality building material "can provide tremendous resilience against earthquake-induced strong ground motions" (pp. 854–855).

Following the 2015 Gorkha earthquake, Ray (2017) observed that structures with lower building heights, built to traditional building designs and using traditional construction technology, survived the severe earthquake. Traditional construction consists of wooden frames, double-framed windows and doors, and timber wedges. Once residents added floors, often brick and cement, or changed the projections to traditional houses, this increased vulnerability, and diminished resilience. He also found that "traditional open spaces and building designs added to the structural resilience" (p. 653).

Following the earthquake, the Planning Commission (2015) identified the building construction technologies likely to be involved in the housing reconstruction process. The options were developed for three building topologies namely:

**Fig. 10** Bahrabise on the East bank of Koshi River

low-strength masonry, cement-based masonry, and reinforced concrete frame for small to medium-sized buildings (p. 14). Low-strength masonry utilizes stone and brick while cement-based masonry utilizes brick or blocks manufactured from concrete or stabilized soil founded on linear wall footings. Reinforced concrete frame buildings should be founded on isolated column footings tied to a foundation beam or raft slab. Attic floors can be stabilized mud on timber joists and planking or reinforced concrete/ reinforced brick slab in the case of low-strength masonry. The upper floor of cement-based masonry or reinforced concrete framed buildings should be reinforced concrete although a reinforced brick floor may be used with cement-based masonry structures. Roofs should be reinforced concrete slabs although the masonry structures may instead utilize corrugated iron sheets or a reinforced brick slab. For masonry buildings the seismic resistant elements for masonry buildings should be reinforced concrete bands, ties, and stitches although wooden or bamboo bands may be utilized in low-strength masonry structures. Reinforced concrete structures must be designed with fully compliant with ductile detailing.

**Fig. 11** Aerial view of typical development in the mountainous areas of Nepal with buildings on the ridges

**Fig. 12** Confined masonry construction as used in Cyprus (September 2022)

The Department of Urban Development and Building Construction has issued several guidelines under the Nepal National Building Code to address earthquake-resistant building construction:

- NBC 203: 2015—Nepal National Building Code: Guidelines for Earthquake Resistant Building Construction: Low Strength Masonry
- NBC 204: 2015—Nepal National Building Code: Guidelines for Earthquake Resistant Building Construction: Earthen Building
- NBC 105: 2019—Commentary: Nepal National Building Code: Seismic Design of Buildings
- NBC 105: 2020—Nepal National Building Code: Seismic Design of Buildings in Nepal

Other laws or bylaws can override the guidelines for earthquake-resistant building construction. Nevertheless, they provide considerable detail on how to find a suitable location based on simple geotechnical rules and then design and build such low-cost structures. They also utilize and describe in detail the traditional methods of building earthquake-resistant buildings. For instance, NBC 204 describes how to prepare non-erodible mud plaster for wattle and daub, mud, and sun-dried brick walls to protect against splash-water erosion, mainly during the wet season. The following ingredients are required to cover over 100 $m^2$ of wall area:

- Mud (1.5 $m^3$)
- Bhusa (chopped wheat or paddy straw) (90 kg)
- 80/100-grade bitumen (85 kg)
- Kerosene oil (16 l)
- Cow-dung (0.1 $m^3$) (NBC 204:2015, p. 54).

While the rural population is resourceful and resilient concerning their housing needs, the local authorities must provide a positive oversight role. They must also ensure that residents can access services and facilities to benefit from development just as the urban population does.

## Roads and Bridges

Accessibility is a critical issue for rural municipalities. A study of six remote rural municipalities in the rugged geographical terrain of Karnali Province found that 65% of the population in two of the municipalities needed to travel for at least 8 hours to the nearest hospital (Walcher et al., 2021, p. 3). Essentially, the population in one municipality must travel at least 8 hours. In the other three municipalities, between 28 and 60% of the population were within 4 hours of the nearest hospital.

Pande (2017) reviewed major rural road projects undertaken in Nepal by multilateral funding institutions and argues that various projects tend to support the view that the contribution and impacts of improved rural connectivity improve the well-being of the rural population (p. 47). He further argues that "[a]s the country's population resides predominantly in the rural areas, there are ample opportunities in promoting and facilitating the improvement of rural transport connectivity and rural–urban linkages to achieve SDGs well ahead of the target date of 2030." If the rural road is "well planned, designed and properly implemented, the improved rural transport connectivity and rural–urban linkage" would provide several significant benefits. These include:

a. Connecting with essential services such as primary, secondary, and tertiary education, health, markets, access to potable water, electricity, and government administration and welfare facilities, which would also offer employment opportunities.
b. Providing employment in the construction and maintenance of rural roads allows rural people to upskill using local technology and local tools.
c. Providing equal opportunities and equal wages to promote gender equality.
d. Facilitating the increased movement of goods and passenger services by the promotion of non-motorized transport.
e. Facilitating the marketing of farm produce; improving horticulture and crop farming by improving the distribution of seeds and development of agricultural cooperatives.
f. Providing transport access for the development of hydropower and other industries (Pande, 2017, p. 47).

Two road sector initiatives are significant and are discussed in detail below.

## Trail Bridges

The Swiss government has funded an annual trail bridge program in Nepal since 2009 (Department of Local Infrastructure, 2022). On average, around 450 bridges are constructed annually. There are currently 8,444 operational trail bridges, and a further 4,000 to 5,000 are still required. The impact of the opening of a trail bridge is significant (Department of Local Infrastructure, 2022; Helvetas, 2022):

a. the routes, serving an average of 1,800 people, are shorter and safer.
b. school attendance increases by an average of 16%.
c. consultations at health centers go up 26%.
d. in 20% of cases, new stores, snack bars, and repair shops open near the bridge.

The Department of Local Infrastructure (2022) estimates that "trail bridges saved lives and 2.3 hours for a two-way journey. Over 1.3 million people cross trail bridges daily to access schools, perform household chores, seek treatment from health facilities and reach markets." The impact of the trail bridge program goes well beyond the provision of access; locals run the program and develop new designs as needs evolve (Helvetas, 2022). Helvetas has been commissioned to provide technical assistance to the program from December 2019 to November 2023. Importantly this assistance "includes working with educational institutes in providing technical courses to engineers and technicians. Women engineers are especially encouraged through an SDC traineeship program" Helvetas (2022). Nepalese engineers are providing advice in other countries, highlighted in an agreement between the Nepalese government and the European Organization for Nuclear Research (CERN) that enables Nepalese engineers and scientists to attend CERN schools and contribute to ongoing research projects (Ministry of Foreign Affairs, Nepal, 2017).

Planning and implementation, including bridge design, have now been devolved to the municipalities with technical backup as required. A bridge user group committee must be established for each short-span bridge (under 120 m). The committee must include women and disadvantaged groups to allow them to "discuss the bridge site, gain an understanding of basic technical issues, organize materials, and contribute hard days of labor (or the financial equivalent). In the process, they make the bridge their own" (Helvetas, 2022). Because men are typically more likely to work in industry and maintain livelihoods in urban contexts, women and disadvantaged groups are often more reliant on rural facilities (International Labour Organization, n.d.). By including the groups most dependent on these bridges in the decision-making process, the results can be more tailored to their most relevant needs. After construction, a public meeting is held to "verify that it meets quality standards and that the funds have been used properly. In addition, a Bridge Warden (caretaker) is appointed to oversee the operation and upkeep of the bridge" (Helvetas, 2022). Corruption has been an ongoing concern in projects within Nepalese municipalities, including overestimation of costs, exchanges of bribery, and preferential design to advantaged groups (Shrestha, 2007). These public meetings, and the inclusion of trustworthy caretakers, can reduce the level of corruption and ensure that the benefits of the project are reaped equally and equitably. For many bridge users this is their first experience of political participation "[w]hich makes each trail bridge an opportunity for local democracy in practice" (Helvetas, 2022).

## Road Maintenance Groups

The *Strengthening the National Rural Transport Programme* (SNRTP) is an initiative of the Government of Nepal and the World Bank with technical support from the International Labour Organization (ILO) (International Labour Organization, n.d., p. 1). While the program's main objective was to maintain and upgrade rural roads and water crossings, the technical focus of the ILO was to promote labor-intensive road maintenance methods to create viable work opportunities (Fig. 13).

**Fig. 13** Road maintenance group in Rupakot

Several associated activities were undertaken to promote gender equality and empower women, including:

a. Developing an RMG guideline to include a target of 100% participation by women and a minimum of 33% with an output-based payment system so that men and women can both adjust work schedules to meet their needs.
b. Prioritizing the enlistment of female engineers.
c. Prioritizing the engagement of female administrative and finance positions so that women can be recruited for more than manual labor.
d. Ensuring equal pay for equal work.
e. Providing specific training on maintenance, bioengineering, and plantation work.
f. Supporting access to financial services.
g. Developing gender-friendly occupational safety and health guidelines.
h. Negotiating free transport to and from the work site.
i. Provide alternative income generation activities for the spouse and family members (International Labour Organization, n.d., pp. 2–3).

Concrete quality of life impacts included: "13% of RMG members renovated their earthquake-affected houses, 34% constructed toilets, 70% became involved in animal husbandry and poultry farming, and 45% are sending their children to a better school thanks to the incomes earned from the road maintenance work" (International Labour Organization, n.d. p. 3). The program also collaborated with a major bank to provide financial services to RMG members (Jha, 2022).

*Guidelines for Road Maintenance Groups* in Nepal have been published by The Department of Local Infrastructure Development and Agricultural Roads (DOLIDAR) (2016).

## Achieving the Sustainable Development Goals

How, then, can the Sustainable Development Goals be achieved in the built environment in such a complex region? The author has developed the matrix in Table 3 to analyze the infrastructure required to meet these goals. There is an apparent link between the necessary human resources and the built environment in which they will live. The responses in the matrix can be objective in some cases and subjective in others but show the built environment's importance in achieving those goals.

**Table 3** The role of infrastructure in achieving the Sustainable Development Goals in rural municipalities

| Sustainable Development Goal (United Nations General Assembly, 2015) | Infrastructure requirements |
|---|---|
| Goal 1. End poverty in all its forms everywhere | This requires an enlightened administration and bureaucracy. Access to quality housing, education, health facilities, and population centers for markets and work. A social safety net for those unable to participate in the workforce. Engagement of the local community in the ongoing construction and maintenance of local infrastructure and ensuring that training and upskilling are required components of the process |
| Goal 2. End hunger, achieve food security and improved nutrition and promote sustainable agriculture | Provide local agricultural and health professionals who can outreach to the community. These professionals will require all of the community infrastructure required to ensure that they will be long-term residents of the local community |
| Goal 3. Ensure healthy lives and promote well-being for all ages | Ready access to quality health facilities; purpose-built health facilities connected to reliable water, power, and sanitation services; fast internet connections and more mobile phone access; quality accommodation for health staff; access to quality to education services for children of health staff; reliable road access to closest regional city/town; access to timely evacuation services for seriously ill patients |
| Goal 4. Ensure inclusive and equitable quality education and promote lifelong learning opportunities for all | Ready access to quality educational facilities; purpose-built school facilities connected to reliable water, power, and sanitation services; fast internet connections and mobile phone access; quality accommodation for school teaching staff; access to quality health services; reliable road access to the closest regional city/town |
| Goal 5. Achieve gender equality and empower all women and girls | As for Goal 1 |

(continued)

**Table 3** (continued)

| Sustainable Development Goal (United Nations General Assembly, 2015) | Infrastructure requirements |
|---|---|
| Goal 6. Ensure availability and sustainable management of water and sanitation for all | Provision of reticulated water to population centers and alternative sustainable potable water supplies to more isolated population areas; provision of appropriate sustainable sanitation |
| Goal 7. Ensure access to affordable, reliable, sustainable and modern energy for all | Continue the program to move the population from gas cookers to electric induction cookers to remove the need to import gas supplies and use abundant hydropower sources instead |
| Goal 8. Promote sustained, inclusive and sustainable economic growth, full and productive employment and decent work for all | As for Goal 1, Goal 3 and Goal 4 |
| Goal 9. Build resilient infrastructure, promote inclusive and sustainable industrialization and foster innovation | |
| Goal 10. Reduce inequality within and among countries | As for Goal 1 |
| Goal 11. Make cities and human settlements inclusive, safe, resilient and sustainable | As for Goal 1, Goal 3 and Goal 4 |
| Goal 12. Ensure sustainable consumption and production patterns | As for Goal 1, Goal 2 and Goal 4 |
| Goal 15. Protect, restore, and promote sustainable use of terrestrial ecosystems, sustainably manage forests, combat desertification, and halt and reverse land degradation and halt biodiversity loss | As for Goal 1, Goal 2, Goal 3, Goal 6, and Goal 7 |
| Goal 16. Promote peaceful and inclusive societies for sustainable development, provide access to justice for all and build effective, accountable and inclusive institutions at all levels | As for Goal 1, Goal 2, Goal 3 and Goal 4 |

## Discussion

This chapter has provided critical insights into the significant challenges facing residents of rural municipalities, especially those in isolated mountain communities. These populations face challenges inconceivable to the citizens of most countries and their resilience is found throughout each generation. Communities outside of the main population centers face the ravages of the annual monsoon season where widespread flooding occurs on the plains of the Terai. In the mountainous regions it results in annual floods in the river valleys, and landslides on the unstable hillslopes. This results in widespread isolation and a lack of access to essential services. In addition, communities are subject to periodic earthquakes due to the instability of the Himalayan region.

However, there are still issues of capacity and governance at the local level. Following the implementation of federalism, the Local Level Restructuring Recommendation Commission recommended the formation of 753 local government organizations (Pradhan, 2019). As a result, six metropolitan cities, 11 sub-metropolitan cities, 276 municipalities, and 460 rural municipalities were formed. The first local government elections in 2017 resulted in the election of around 37,000 representatives (Pradhan, 2019). Because of gender and social equity provisions in the Constitution, around 15,000 representatives were women and/or Dalits.[9] Skill levels across the various local government units will be quite variable. With such a large number of local government units, competition for qualified staff will continue to be fierce. In addition, there will be a continuing need for basic training of elected representatives and technical and administrative staff. As mentioned earlier, staff retention will require good access to facilities, working conditions, and housing.

Basnyat (2019) considers that one of the early critical achievements at the local level is that they have become the first point of contact between citizens and the state. He found it particularly significant that officials responsible for building inspections and construction permits are now at the ward level. Similarly, complaints such as those challenging assessments and grant-beneficiary lists can be registered at the ward level, whereas once they had to be lodged at the district headquarters. Easy access to local support is essential when local communities are recovering from earthquakes, landslides, and flooding.

The impact of institutional changes that occurred with devolution under the new constitution, and listed as achievements by Basnyat, above, should not be underestimated. Provided that adequate staffing is provided at the local level this

[9] Dalits are the lowest group in the Hindu-based caste system.

puts the local communities in touch with their local politicians and government services. As many of the services are provided at the local (Ward) level, villagers will no longer be required to travel long distances, often on roads and tracks in poor condition, to access these services. It also means that planning and building is controlled to a degree at the local level, ensuring that it is in keeping with the local geotechnical and environmental regime. As has been shown throughout this paper, Nepali people have been very resilient under often harsh conditions. To survive, they have relied on traditional knowledge passed down through generations, much of which is still relevant today, especially at the local level.

## Conclusion

In summary, to achieve the SDGs in the built environment, rural municipalities in Nepal must provide:

a. All-weather access to education, health services, social services, government services, and markets.
b. Quality earthquake-resilient public buildings provided with potable water, sanitation, and electricity.
c. Building standards for domestic residences as well as providing access to potable water, sanitation, and electricity.
d. Reliable internet and mobile phone networks.
e. Schools, hospitals, health centers, and other facilities are of such a standard that staff will be readily recruited and motivated to stay in the municipality. This also requires the provision of quality housing for the families of all members of staff.

In addition, they must ensure that:

a. No significant facilities are constructed without a geotechnical investigation to ensure that the ground is stable and avoids potential earthquake/mudslide locations.
b. As far as possible, both men and women are provided with appropriate training and then employed in the construction, operation, and maintenance processes.

Finally, to achieve equitable outcomes, all three tiers of government must ensure an equitable allocation of funds to all rural communities.

## References

Acharya, K. K. (2018). The capacity of local governments in Nepal: From government to governance and governability? *Asia Pacific Journal of Public Administration, 40*(3), 186–197. https://doi.org/10.1080/23276665.2018.1525842

Acharya, K. K., & Zafarullah, H. (2020). Institutionalizing federalism in Nepal: Operationalizing obstacles, procrastinated progress. *Public Administration and Policy, 23*(2), 125–139. https://doi.org/10.1080/PAP-03-2020-0013

Adhikari, B., & Upadhyaya, P. (2020). Inter-government relations bill: A mixed bag of optimism for Subnational Governments. *New Spotlight, 14*(2). https://www.spotlightnepal.com/2020/07/17/inter-government-relations-bill-mixed-bag-optimism-subnational-governments/

Ahmad, N., Ali, Q., Ashraf, M., Naeem Khan, A., & Alam, B. (2012). Performance assessment of low–rise confined masonry structures for earthquake induced ground motions. *International Journal of Civil & Structural Engineering, 2*(3), 851–868. https://doi.org/10.6088/ijcser.00202030013

Basnyat, B. (2019, August 14). Political transition and earthquake recovery in Nepal. *InAsia.* https://asiafoundation.org/2019/08/14/political-transition-and-earthquake-recovery-in-nepal/

Bell, R., Fort, M., Götz, J., Bernsteiner, H., Andermann, C., Etzlstorfer, J., Posch, E., Gurung, N., & Gurung, S. (2021). Major geomorphic events and natural hazards during monsoonal precipitation 2018 in the Kali Gandaki Valley, Nepal Himalaya. *Geomorphology, 372*(107451), 1–25. https://doi.org/10.1016/j.geomorph.2020.107451

Chaudhary, D. (2019). The decentralization, devolution and local governance practices in Nepal: The emerging challenges and concerns. *Journal of Political Science, XIX*, 43–64.https://doi.org/10.3126/jps.v19i0.26698

*Constitution of Nepal.* (2015). https://lawcommission.gov.np/en/?lsvr_document_cat=existing-law-constitution

Cowan, S. (2021, August 21). The Araniko Highway conundrum. *The Record.* https://www.recordnepal.com/the-araniko-highway-conundrum

Dahal, R. K. (2010). *Engineering geology of Nepal.* Published in personal home page www.ranjan.net.np

Department of Local Infrastructure. (2022). *Trail Bridge Sector Wide Approach, TB SWAp- Framework-III.* https://doli.gov.np/doligov/program-projects/trail-bridge-sector-wide-approach-tb-swap-framework-iii/

Department of Local Infrastructure Development and Agricultural Roads (DOLIDAR). (2016). *Road Maintenance Groups (RMG): Guidelines.* https://www.ilo.org/wcmsp5/groups/public/---asia/---ro-bangkok/---ilo-kathmandu/documents/webpage/wcms_562124.pdf

Devkota, S., Ghimire, S., & Upadhyay, M. (2021). What factors in Nepal account for the rural-urban discrepancy in human capital? Evidence from household survey data. *Economies, 9*(83), 1–14. https://doi.org/10.3390/economies9020083

Dey, S. (2015). A devastating disaster: A case study of Nepal earthquake and its impact on human beings. *IOSR Journal Of Humanities and Social Science, 20*(7), 28–34. https://www.iosrjournals.org/iosr-jhss/papers/Vol20-issue7/Version-5/E020752834.pdf

Gurung, L. (2021, September 6). A brief note on the Education Levels of Nepal. *Blog Post, Edu Sanjal Pvt. Ltd.* https://edusanjal.com/blog/education-levels-nepal/

Helvetas. (2022). *Trail bridges for safe routes*. https://www.helvetas.org/en/switzerland/what-we-do/how-we-work/our-projects/asia/nepal/nepal-infrastructure-suspension-bridge

International Labour Organization. (n.d.). *Nepal: Road maintenance as a vehicle for social inclusion and decent work for women*. https://www.ilo.org/dyn/asist/docs/F354446576/GenderBriefNepal.pdf

Jha, S. K. (2022, March 10). Paving the road to financial inclusion for rural workers in Nepal. *International Labour Organization*. https://www.ilo.org/empent/areas/social-finance/news/WCMS_839334/lang--en/index.htm

Khabar, M. (2022, January 26). These are the most and least populated municipalities. *Makalukhabar*. https://english.makalukhabar.com/2022/01/42069/

Malalgoda, C., Amaratunga, D., & Haigh, R. (2014). Challenges in creating a disaster resilient built environment. *Procedia Economics and Finance, 18*, 736–744. https://www.sciencedirect.com/science/article/pii/S2212567114009976

Ministry of Foreign Affairs, Nepal. (2017). *Press release issued by permanent mission of Nepal, Geneva regarding International Cooperation Agreement between Nepal and CERN*. https://mofa.gov.np/press-release-issued-by-permanent-mission-of-nepal-geneva-regarding-international-cooperation-agreement-between-nepal-and-cern/

Mulmi, D. A. (2009). Green road approach in rural road construction for the sustainable development of Nepal. *Journal of Sustainable Development, 2*(3), 149–165. https://doi.org/10.5539/jsd.v2n3p149

Muñoz-Torrero Manchado, A., Allen, S., Ballesteros-Cánovas, J. A., Dhakal, A., Dhital, M. R., & Stoffel, M. (2021). Three decades of landslide activity in western Nepal: New insights into trends and climate drivers. *Landslides, 18*, 2001. https://doi.org/10.1007/s10346-021-01632-6

National Planning Commission. (2015). *Nepal Earthquake 2015: Post Disaster Needs Assessment. Vol. B Sector Reports*. https://nepal.unfpa.org/sites/default/files/pub-pdf/PDNA_volume_BFinalVersion.pdf

*NBC 203: 2015 - Nepal National Building Code: Guidelines for earthquake resistant building construction: Low strength masonry*. https://www.dudbc.gov.np/uploads/default/files/c89e75cec2dc78d9973d92c3897fd71a.pdf

*NBC 204: 2015 - Nepal National Building Code: Guidelines for Earthquake resistant building construction: Earthern building*. https://www.dudbc.gov.np/uploads/default/files/52e69c2ecf9c35bb64a8f22884dc4f8c.pdf

*NBC 105: 2019 - Commentary: Nepal National Building Code: Seismic design of buildings*. https://www.dudbc.gov.np/uploads/default/files/ca89c3c4a62d87354a2cd162746ba4df.pdf

*NBC 105: 2020 - Nepal National Building Code: Seismic design of buildings in Nepal*. https://www.dudbc.gov.np/uploads/default/files/9a192ea8b7e1c45b99628f0869052201.pdf

"Nepal Earthquake Case Studies: Engineering, Infrastructure, Design". (2015). [Student Report]. *Nepal earthquake case studies: Engineering, infrastructure, design*. Dartmouth University. https://journeys.dartmouth.edu/NepalQuake-CaseStudies/engineering-infrastructure-design/

Pande, K. (2017, March 14–16). *Improving rural connectivity and its impact on Sustainable Development Goals (SDGs)—Case of Nepal: Final report*. Intergovernmental Tenth

Regional Environmentally Sustainable Transport (EST) Forum in Asia, Vientiane, Lao People's Democratic Republic. https://www.uncrd.or.jp/content/documents/4937Final%20Programme%20-Lao%20EST%20Forum%20-16March2017.pdf

Paudel, B., & Sapkota, K. P. (2018). *Local levels in federalism: Constitutional provisions and the state of implementation.* Swatantra Nagarik Sanjal. https://asiafoundation.org/wp-content/uploads/2018/08/Local-Levels-in-Federalism_Constitutional-Provisions-and-the-State-of-Implementation-English-Version.pdf

Poudel, A. (2022, June 10). No lesson learned from the last year's monsoon devastations. *Kathmandu Post*. https://kathmandupost.com/climate-environment/2022/06/02/no-lesson-learned-from-the-last-year-s-monsoon-devastations

Pradhan, K. M. (2019). Local government: Constitutional vision and present practice. *Research Nepal Journal of Development Studies, 2*(2), 109–128. https://doi.org/10.3126/rnjds.v2i2.29284

Purdie, E., Diehl, A., & Iimi, A. (2016, September 12). On the road to sustainable growth: measuring access for rural populations. *Data Blog*. https://blogs.worldbank.org/opendata/road-sustainable-growth-measuring-access-rural-populations-0

Ray, B. (2017). Response of a resilient community to natural disasters: The Gorkha Earthquake in Nepal, 2015. *The Professional Geographer, 69*(4), 644–654. https://doi.org/10.1080/00330124.2017.1298452

Sapkota, J. B. (2018). Access to infrastructure and human well-being: Evidence from rural Nepal. *Development in Practice, 28*(2), 182–194. https://doi.org/10.1080/09614524.2018.1424802

Shrestha, P. M. (2007). *Corruption in infrastructure provision and service delivery at the municipal level in Nepal: Case study*. WEDC Loughborough University UK. https://assets.publishing.service.gov.uk/media/57a08c0ced915d622c0010bf/WEDC-Nepal-casestudy.pdf

Shrestha, M. S., & Mishra, A. (2020, October 13). Better governance is a key to reducing flood risks. *myRepublica*. https://myrepublica.nagariknetwork.com/news/better-governance-is-a-key-to-reducing-flood-risks/?categoryId=opinion

Smith, R. B., & Smith, N. N. (2022). Federalism and regional autonomy: Overcoming legal and administrative hurdles in Nepal. *Lex Scientia Law Review, 6*(2), 223–264. https://doi.org/10.15294/lesrev.v6i1.54437

Suvedi, M. (2010). Building capacity for sustainable rural development: Lessons from Nepal. *International Journal of Environmental and Rural Development, 1*(1), 188–193. https://iserd.net/ijerd11/11188.pdf

United Nations General Assembly. (2015). *Resolution A/RES/70/1: Transforming our world: the 2030 Agenda for Sustainable Development*. https://www.un.org/ga/search/view_doc.asp?symbol=A/RES/70/1&Lang=E

United Nations General Assembly. (2021). *Resolution A/76/L.6/Rev.*1: Graduation *of Bangladesh, the Lao People's Democratic Republic, and Nepal from the least developed country category*. https://www.un.org/ldcportal/news/general-assembly-issues-resolution-graduation-bangladesh-lao-pdr-and-nepal

Upreti, B. N. (1999). An overview of the stratigraphy and tectonics of the Nepal Himalaya. *Journal of Asian Earth Sciences, 17*(5–6), 577–606. https://doi.org/10.1016/S1367-9120(99)00047-4

Valagussa, A., Frattini, P., Valbuzzi, E., & Crosta, G. B. (2021). Role of landslides on the volume balance of the Nepal 2015 earthquake sequence. *Scientific Reports, 11*(3434), n.p. https://www.nature.com/articles/s41598-021-83037-y

Walcher, A., Adhikary, P., Gosain, B., Tamang, U., Uematsu, H., Banick, R., Maharjan, L., & Shrestha, T. (2021). Bridging the gap between data and policies in remote rural municipalities in Nepal. *Property & Equity Notes,* (39), 1–5. https://openknowledge.worldbank.org/bitstream/handle/10986/35201/Bridging-the-Gap-Between-Data-and-Policies-in-Remote-Rural-Municipalities-in-Nepal.pdf?sequence=5

Wikiwand. (2021, January 18). *List of gaunpalikas of Nepal.* https://www.wikiwand.com/en/List_of_gaunpalikas_of_Nepal

World Bank. (2022). DataBank: Health nutrition and population statistics. https://databank.worldbank.org/source/health-nutrition-and-population-statistics

# Correction to: Determining an Adequate Population Density to Achieve Sustainable Development and Quality of Life

Ahmed Salem

**Correction to:**
**Chapter "Determining an Adequate Population Density to Achieve Sustainable Development and Quality of Life": T. Walker et al. (eds.), *The Role of Design, Construction, and Real Estate in Advancing the Sustainable Development Goals*, Sustainable Development Goals Series, https://doi.org/10.1007/978-3-031-28739-8_4**

The original version of Chapter 4 was inadvertently published with an incorrect word 'Low Density' in Figure 11, which has now been removed. The chapter and the book have been updated with the changes.

The updated original version of this chapter can be found at https://doi.org/10.1007/978-3-031-28739-8_4



T. Walker et al. (eds.), *The Role of Design, Construction, and Real Estate in Advancing the Sustainable Development Goals*, Sustainable Development Goals Series, https://doi.org/10.1007/978-3-031-28739-8_12

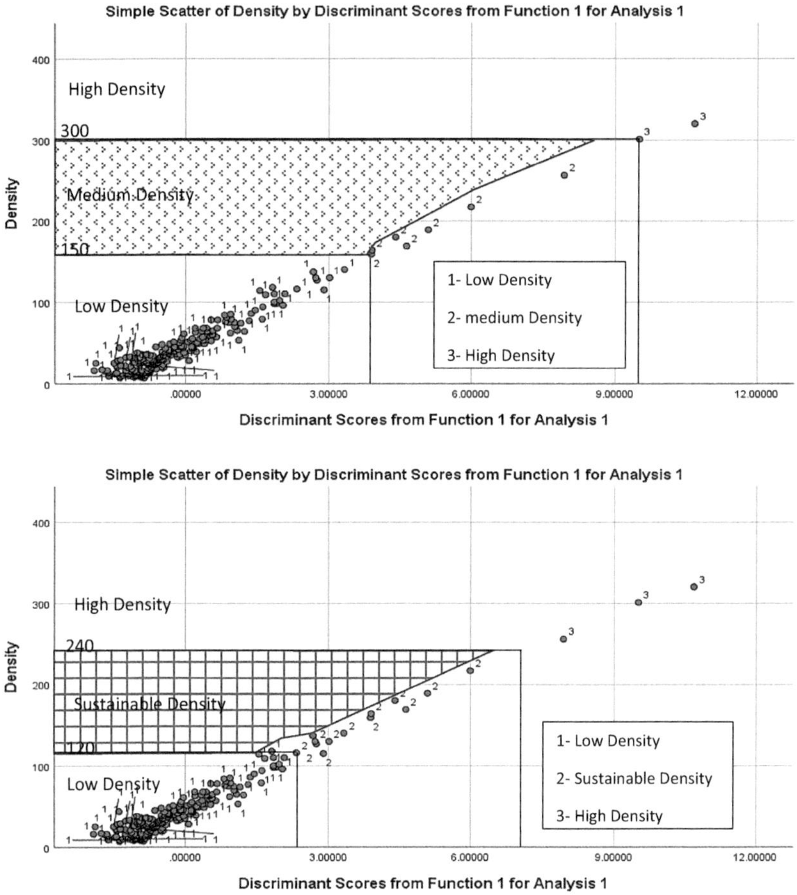


**Fig. 11** A comparison between the Tonkin study and this study to determine density classification and sustainable density (author-originated)

# Index




T. Walker et al. (eds.), *The Role of Design, Construction, and Real Estate in Advancing the Sustainable Development Goals*, Sustainable Development Goals Series, https://doi.org/10.1007/978-3-031-28739-8

GPSR Compliance

*The European Union's (EU) General Product Safety Regulation (GPSR) is a set of rules that requires consumer products to be safe and our obligations to ensure this.*

*If you have any concerns about our products, you can contact us on ProductSafety@springernature.com*

In case Publisher is established outside the EU, the EU authorized representative is:

Springer Nature Customer Service Center GmbH
Europaplatz 3
69115 Heidelberg, Germany

Batch number: 10732017

Printed by Printforce, the Netherlands